CIRCADIAN
PHYSIOLOGY

CIRCADIAN PHYSIOLOGY

Roberto Refinetti, Ph.D.

CRC Press

Boca Raton   London   New York   Washington, D.C.

Library of Congress Cataloging-in-Publication Data

Refinetti, Roberto.
    Circadian physiology / Roberto Refinetti.
       p. ; cm.
    Includes bibliographical references and index.
    ISBN 0-8493-2299-5 (alk. paper)
    1. Circadian rhythms. I. Title.
    [DNLM: 1. Circadian Rhythm—physiology. QT 167 R332c 1999]
  QP84.6 .R534 1999
  571.7'7—dc21

                99-051743
                CIP

No claim to original U.S. Government works
International Standard Book Number 0-8493-2299-5
Library of Congress Card Number 99-051743
Printed in the United States of America 1 2 3 4 5 6 7 8 9 0
Printed on acid-free paper

Preface

The idea of writing this book came to me 6 years ago, when I started teaching an upper-level course on circadian rhythms and was unable to find a suitable textbook. The available books were either too limited in scope (addressing specialized topics in the field) or too old and out-of-date. Teaching and research responsibilities prevented me from writing the book until very recently, when I took a year off from academia and found enough time to go over the extensive literature in the field and organize my own thoughts.

The need for a textbook remains unchanged today, but I now realize that a book on circadian physiology is just as badly needed for a wide audience of biologists, physiologists, neuroscientists, and psychologists who must deal with circadian processes in their specialized research fields but who have not had a formal education in biological rhythms. A great deal of our current knowledge on biological rhythms has been acquired during the last 30 years and was not available to most contemporary researchers when they were in graduate school, much less when they were undergraduate students. Therefore, an authoritative, up-to-date review of scientific and medical advances in circadian biology accessible to non-specialists is urgently needed. To provide the in-depth understanding of circadian phenomena required for the analysis of actual research data, I included in this book a series of computer programs and guided exercises that will allow readers to put into practice the formal knowledge acquired through the disciplinary chapters. In addition to providing an entrance route to circadian physiology for life-science researchers, this book should also be valuable as a textbook for graduate and advanced undergraduate instruction.

Because of the recent growth in the exploitation of circadian topics in the popular literature, I felt that it would be appropriate to start the book with a brief account of how biomedical scientists distinguish facts from superstition. I follow this introductory chapter with one that discusses some specific misuses of concepts in biological timing. In the third chapter, I present a general review of biological rhythms to properly situate the field of circadian physiology, which I review in detail in the following seven chapters. The final four disciplinary chapters are dedicated to the application of knowledge on circadian rhythms to business and health care issues.

Author

Roberto Refinetti, Ph.D., is a physiological psychology professor and an active researcher in the field of biological rhythms. He has published more than 100 articles in professional journals and edited several books. He has been award-ed research grants by the National Science Foundation and the National Institutes of Health and has conducted research at the University of Califor-nia–Santa Barbara, the University of São Paulo, The University of Illinois, the University of Virginia, and the College of William & Mary. At present, he is a member of the faculty at the University of South Carolina–Salkehatchie. His web site is www.circadian.org.

Acknowledgments

I would like to thank CRC Press for agreeing to publish this book. Being accustomed to publishing research articles in professional journals, I did not anticipate the difficulties involved in publishing a book. The editorial staff at CRC Press, particularly Barbara Norwitz and Sylvia Wood, was extremely helpful. I am equally grateful for the support of my research in circadian physiology provided by grants from the National Science Foundation and the National Institutes of Health. Also, the collaboration of several students of mine was of great importance for my research enterprise. In particular, I would like to acknowledge the contributions by Aaron Osborne, Candice Brown, Adam Shoemaker, Randy Watts, Steve Susalka, and Adam Daymude.

I entered the field of circadian physiology in the late 1980s as a postdoctoral associate in Dr. Evelyn Satinoff's laboratory at the University of Illinois and in Dr. Michael Menaker's laboratory at the University of Virginia. Their contribution to my professional development is greatly appreciated. Further back in time, I would not even have started a career in behavioral physiology were it not for the support and encouragement of my graduate advisors in the early 1980s, Dr. Dora Ventura at the University of São Paulo, and Dr. Harry Carlisle at the University of California, Santa Barbara. To my daughter, Lauren Refinetti, who at the age of 7 attended my classes on circadian physiology and provided helpful feedback, I also owe a debt of gratitude. "Caring and sharing" from my fiancée, Jennifer Palmer, helped me move through piles of reprints and books unharmed. Finally, my professional and personal life would not have been possible without the solid support from my parents throughout the years.

Dedication

This book is dedicated to my father, in memoriam.

Contents

part I

---

*the biological clock:
facts and hoopla*

chapter one

facts and superstition

> To help readers differentiate between serious and far-
> cical books on biological clocks, this chapter briefly
> explains the basic principles used by scientists to sep-
> arate fact from superstition.

Libraries and bookstores contain hundreds of books dealing with "biological clocks," "biological rhythms," "biorhythms," and other topics related to the timing of biological processes. Not surprisingly, some of these books present reliable facts, while others border on superstition and nonsense. How can a prospective reader tell them apart? The author's credentials and the repu-tation of the publisher are good clues, but any serious reader would like to be able to use his or her own judgment to make a decision. Is there a simple but reliable procedure to separate fact from hoopla?

Let us start with a simple test. For each of the following six statements give a "true or false" judgment:

True or False
---

1. The capital of the United States is Washington, D.C.
2. Delaware is twice the size of Texas.
3. All humans are mortal.
4. All rabbits are gray.
5. All ziwats are blue.
6. Nasty people are possessed by an evil spirit that cannot
   be detected by any known means.

I realize that these statements do not have the same logical structure, but I hope that this did not prevent you from trying to answer them. State-ments 1, 2 , and 3 were quite easy. All three of them referred to well-known facts. Statement 1 is *true* because Washington has been the capital of the United States since the year 1800. Statement 2 is *false* because Delaware is much smaller than Texas. Statement 3 is *true* because no human being, out of the billions who lived before us, has ever lived past the age of 200 (the current life expectancy at birth in the United States is only 76 years).[1] It is evident that, when the contents of a statement are well established, it is easy to judge its veracity.

Statement 4 was a little less obvious. The statement is *false* because, although some rabbits are gray, not all rabbits are gray (some are white, for instance). Statement 5 was even less obvious. As a matter of fact, it could not be answered at all. Since we do not know what "ziwats" are, we cannot attest to whether they are all blue or not. It should be noticed, however, that the statement is testable: once we know what ziwats are, we can look at various ziwats and, if we find a ziwat that is not blue, we can safely conclude that Statement 5 is *false*. Thus, if "ziwats" is a synonym of "birds," then we can easily conclude that the statement is false as soon as we see a cardinal or a sparrow flying by.

Statement 6 was different in nature from the others. Although it contained two sub-statements, this is not where the main difference resides. The main difference resides in the impossibility to test the first sub-statement. Because, according to the second sub-statement, evil spirits cannot be detected by any known means, it is impossible to determine the veracity of the first sub-statement (i.e., whether nasty people are possessed by an evil spirit or not). Statement 6 may be *true* or may be *false* — we have no way of knowing which is the case. Only superstition (or faith, or whatever one chooses to call it) would justify accepting the first statement as *true*. If anything, caution would incline us to say that the statement is *false*.

So, there seems to be a simple way to tell fact from superstition. Philosophers of the "contructivist" school have made it abundantly clear that scientific knowledge may be as fallible and incomplete as any product of human activity,[2] including superstitious thought. However, scientific uncertainties are very different from superstitious assumptions. A scientific uncertainty is a hypothesis waiting to be tested, whereas a superstitious assumption is just that — an assumption. Superstitions are "facts" accepted without question; real facts are scientific hypotheses that have been put to test.

Take the case of horoscopes. There is no scientific evidence whatsoever that the alignment of planets in the solar system has an effect on the behavior of human beings on earth (except, of course, that awareness of the alignment causes superstitious people to behave differently). Yet, hundreds of thousands of people read their horoscopes each day. Some do it just for entertainment, but many do it because they believe that the particular alignment of planets will actually affect their daily lives. This is superstition at its best. Why? Because the hypothesis of a link between planet alignment and human predisposition is never put to test. By chance (if not by suggestion), some of the predictions of the horoscope do turn out to be true once in a while, and the credulous reader seems to be satisfied with that. Should the credulous reader bother to write down all the predictions each day and have somebody count how many predictions turned out to be correct and how many turned out to be wrong, he or she would soon realize that the predictions are no better than random guessing.

Let us say that all the astrologer had to do would be to predict whether the horoscope reader will have a good day or a bad day. Obviously, the odds that the astrologer would guess correctly 100 times out of 100 are very small — but even the credulous reader does not expect such an outstanding performance. What about 10 out of 10? The odds in this case are not that small, but the situation is still unrealistic. More likely, the astrologer would predict correctly perhaps 7 out of 10 times. The credulous reader would consider this outcome very good and would not even bother to consider the odds of this outcome happening by chance (that is, by pure guessing). In actuality, standard statistical procedures allow us to calculate that the odds are higher than 1 in 20 (and I doubt this same person would ever drive on a highway if the odds of having a fatal accident were 1 in 20). Thus, if the horoscope reader would seriously put to test the hypothesis of a causal link between planet alignment and human predisposition, he or she would refute the hypothesis. Yet, just like the compulsive gambler who remembers the winning streaks much better than the devastating losses, the horoscope reader refuses to face reality — in both cases, superstitious behavior is at work. Now, indulging in superstitious behavior may even be fun sometimes, but the fun does not make it scientific.

To use the terminology of the late logician Sir Karl Popper,[3] scientific statements are refutable (although not actually refuted) by experimentation. Statement 4 ("All rabbits are gray") is a scientific hypothesis that was put to test and was refuted by the observation that there are white rabbits. Statement 3 ("All humans are mortal") is a scientific hypothesis that has been put to the test and has never been refuted — so many times has it not been refuted that we consider it a scientific fact. Assuming that ziwats actually exist, Statement 5 ("All ziwats are blue") is also a scientific hypothesis, which can be easily refuted by the sighting of a single red or gray ziwat. On the other hand, Statement 6 ("Nasty people are possessed by an evil spirit that cannot be detected by any known means") cannot be tested and, therefore, has no scientific value. When properly placed, it could provide good entertainment in a fiction book; but it has no place in the non-fiction world.

Of course, superstition has similarities with faith — as in religious faith. Because most non-scientists (and many scientists as well) have a religious faith, it is important to ask whether religions can be scientific and whether scientists must be atheists. This question deserves a careful answer. On the one hand, traditional religions have no interest in being scientific. They encourage man's search for knowledge about the world, but their mission is of a moral nature. Religious faith is not scientific, but it does not wish to be. As a matter of fact, accepting a religious doctrine against empirical evidence is a demonstration of greater faith (and, therefore, of greater religiosity). On the other hand, dogmatic faith is incompatible with scientific inquiry, and it has been argued that religious education is detrimental to science education.[4] However, this does not mean that scientists must be atheists. A religious scientist can appeal to faith in religious matters but resort to scientific procedures in empirical matters. The two perspectives can coexist peacefully in the same individual. In a few instances — such as the conflict between creationism and evolution for members of some Protestant denominations — religious faith requires abandonment of the scientific approach. In these cases, religious and scientific interests do collide. It is then up to the individual to choose the path to follow. In the religious perspective, evolution must be wrong because it contradicts the faith on creationism. In the perspective of modern biology, creationism is nothing more than a historical superstition. As far as life on earth is concerned, science is better than faith in terms of knowledge acquisition; however, if one believes in life after death (a truly non-scientific faith), then faith may be more important than science in many situations.

Scholars have argued that scientific literacy has been in a steady decline in America[5] and that this illiteracy is responsible for an increasing acceptance of superstition among the lay public and academics as well.[6] We have just seen, however, that the ability to distinguish fact from superstition requires no more than simple commonsense logic. Of course, even the most sensible reader can be misled by the presentation of wrong facts, but no one is expected to be a specialist on every topic about which he or she reads. When it comes to the facts themselves (i.e., the presentation of the results of scientific tests), the reader should be able to trust the author. Fraud in scientific research is a serious matter, but the scientific community is well aware of it and takes the necessary steps to prevent its incorporation into the accepted body of knowledge.[7] What is sometimes confusing is the ambiguity in experimental results, which allows different authors to paint different pictures of the same event.

Ambiguity in experimental results? Yes. Although non-scientists often think of science as the domain of certainties, ambiguities are at the very heart of scientific research.[8] Consider the statement, "Cigarette smoking causes lung cancer." After decades of research on the effects of smoke inhalation and nicotine consumption, it is generally agreed today that cigarette smoking does cause lung cancer. However, claims of this causal relationship were being made well before laboratory evidence was available. Many people accepted the veracity of the relationship based solely on one of the first epidemiological findings, namely, that people who smoke cigarettes are more likely to develop lung cancer. What is wrong with this reasoning? A lot. Researchers have no control over who smokes and who does not. For multiple unknown reasons, some people become smokers and some do not. This means that it could very well happen that smokers share some gene complex that inclines them to both smoke and contract cancer. If this were the case, they would contract cancer whether or not they smoked. If we could tell who has this gene complex and who does not, and if we could convince some of the people who have the gene not to smoke, then we could have an unequivocal answer. But we cannot do any of this. Consequently, the epidemiological data alone cannot establish a causal link between smoking and cancer. In the statistician's jargon, this is described by the statement that correlation does not imply causation. Of course, parallel research has provided evidence that the correlation between smoking and cancer does reflect causation. However, in the absence of this additional evidence, the correlation provided only ambiguous results. This is especially evident if we consider an epidemiological finding that does not have the support of independent laboratory data.

People who pay attention to their car insurance policy know that premiums are lower for married individuals than for single ones. This is a consequence of the "epidemiological" fact that married drivers are less likely to have car accidents. So, there is a correlation between marriage status and car accidents. Does this mean that getting a marriage license makes you a more responsible driver? Definitely not. What probably happens (and I am unaware of actual studies on this) is that people who value the institution of marriage are also more circumspect in their habits, including their driving. The correlation between marriage status and car accidents is not due to a causal link between the two variables but to the action of a third factor. The same could have been true about the correlation between smoking and cancer. It just happened not to be so.

The distinction between correlation and causation is only one source of ambiguity in scientific research, but it exemplifies the origin of conflicting descriptions of scientific findings to the general public. Let us say that two researchers identified the correlation between smoking and cancer many years ago and wrote books about it. One of them, very concerned about how many lives could be saved if the correlation involved causation, alerted her readers about the dangers of smoking. The other, greatly concerned about scientific accuracy, severely criticized his colleague's book. Who should the readers have believed? Or let us take a more radical example: It is an epidemiological fact that people with more education have sex less often.[9] Does this mean that going to college reduces one's sexual capability? Most likely not. People with more education probably develop interests that compete for time with sexual activities. But, based on the correlation alone, no one can tell.

What can the reader do about this ambiguity? My answer is that it depends on how desperate one is. If something sounds too good to be true, it probably is — but hope sometimes is more important than the facts themselves. A cancer patient may very well accept a questionable new treatment as true because he or she has nothing to lose. Someone with an occasional headache will be much more suspicious about a new miraculous pain killer. Those who can afford to look at science from a purely intellectual perspective will want to watch out for ambiguities and will demand rigorous experimental control of the variables under study.

Rigorous experimental control is no esoteric concept. It is the simple application of the very intuitive principle of determinism. This principle can be stated simply as "Every effect has a cause." The water for your coffee will not boil unless you light the stove (or provide heat by means of an electric heater, a microwave oven, etc.). A female dog will not become pregnant unless she has sex with a male dog (or is artificially inseminated). You will not get to your in-laws' house unless you drive there (or walk, or fly, etc.). It is true that the principle of determinism does not apply in full to the most fundamental level of reality involved in quantum mechanics, but it does apply to the remaining levels of reality.[10] Of course, the principle of determinism itself cannot be experimentally demonstrated, as it is a prerequisite of the experimental method. It is, as postmodernists often point out, an article of the scientist's faith.

Naturally, an effect may have more than one cause. This is why every serious research involves a control group. If you want to find out whether sex causes pregnancy, it is not enough to pair, say, six male dogs with six female dogs; you must also have six female dogs that are not paired with males. The fact that the six paired dogs get pregnant, while none of the unpaired ones do, allows you to conclude that sex is the cause of pregnancy. Without the control group, you would not be able to exclude an infinite number of alternative explanations, such as "the spirit of pregnancy fell upon the female dogs at the same time as they were paired with the males." If the "spirit of pregnancy" did fall on the dogs, it should have fallen on all 12 dogs, not just on the six paired ones. Of course, the pairing with a male (without actual sex) could be the stimulus for the fall of the spirit. If you had this suspicion, your control group should consist of female dogs paired with infertile male dogs. Clearly, the idea is to have a control group that is identical to the experimental group except for the element that you are studying. Since the control group is not exposed to the cause, it does not display the effect. No cause, no effect. No heat, no hot coffee.

In clinical studies, experimental control involves one additional element: the placebo control. Placebo, as you probably know, refers to fake medication — sugar pills intended to please the patient without having any real pharmacological effect. There is much more to it, though.[11] It has long been known that placebo medication actually improves the condition of a small but significant number of patients. This means that the mere belief that one is receiving adequate medication may improve one's condition. It is a "psychological" cure, which means that some unknown process in the brain has the same effect on, say, the heart as the intended drug has. Consequently, if we want to know what the effects of the actual drug are, it is not enough to have an experimental group and a control group; a placebo group is also needed. Thus, any changes observed in the control group will be due to natural remission rates (yes, sometimes illnesses disappear on their own, as when a headache goes away even though you did not take any pain killer); changes in the placebo group will be due to the combined effect of natural remission and "psychological" processes; and changes in the experimental group will be the result of the combination of natural remission, psychological processes, and the specific effect of the drug. Thus, if the results indicate improvement in 10% of the patients in the control group, 20% in the placebo group, and 70% in the experimental group, we can conclude that 50% of patients will benefit from the medication (that is, 70% – 20% = 50%). We can also conclude that 10% of the patients will show natural remission and 30% will remain miserable regardless of what we do. By simple subtraction, we could even conclude that 10% of the patients will show psychological remission (that is, 20% – 10% = 10%), although there could be an interaction between natural and psychological remission processes, which would result in non-additivity of the effects.

Of course, not everyone can be a researcher. Extensive knowledge of the literature in a field and practical experience in the design and execution of experiments allow scientists to evaluate the merits of a study better than ordinary people. However, blatant cases of pseudo-science — such as those found in the popular media — can be detected by any serious reader armed with the two concepts that we have just discussed: the distinction between scientific and superstitious reasoning, and the principle of determinism. Examples of this are presented in the following chapter, where I demystify the use of the term biological clock in connection with biorhythms, pregnancy predestination, and aging. The real biological clock has very little to do with biorhythms, with the urge to have children, or even with aging.

bibliographic notes

1. Brunner, B. (1998). *Time Almanac.* Boston: Information Please.
2. Refinetti, R. (1997). Philosophy of science and physiology education. *American Journal of Physiology* 272: S31-S35.
3. Popper, K. (1959). *The Logic of Scientific Discovery.* New York: Harper & Row.
4. Mahner, M. and Bunge, M. (1996). Is religious education compatible with science education? *Science and Education* 5: 101-123.
5. Hirsch, E. D., Jr. (1988). *Cultural Literacy: What Every American Needs to Know.* New York: Vintage.
6. Gross, P. R. and Levitt, N. (1994). *Higher Superstition: The Academic Left and Its Quarrels with Science.* Baltimore: Johns Hopkins University Press.
7. Alberts, B. and Shine, K. (1994). Scientists and the integrity of research. *Science* 266: 1660-1661.
8. Grinnell, F. (1996). Ambiguity in the practice of science. *Science* 272: 333.
9. Laumann, E. O., Gagnon, J. H., Michael, R. T., and Michaels, S. (1994). *The Social Organization of Sexuality: Sexual Practices in the United States.* Chicago: University of Chicago Press.
10. Mautner, T. (1997). *Dictionary of Philosophy.* New York: Penguin.
11. Bunk, S. (1998). Ethical debate on placebo use may prompt new trial designs. *Scientist* 12(8): 1, 7, 14.

chapter two

biorhythms, pregnancy, predestination, and aging

> This chapter shows that the concept of biorhythm is a farce, that the use of the term biological clock to refer to a woman's urge to have children is inaccurate, and that the process of aging is not a biological rhythm.

The expression *biological clock* is sometimes used in connection with biorhythms, pregnancy predestination, and aging. This usage is incorrect. Before we discuss the *real* biological clock, we must expose the misuse of the term.

biorhythms

What is usually referred to as biorhythms is a farce. "Hoax" would perhaps be a more dramatic term, but it implies intent to deceive, for which I do not have conclusive evidence. As described in Chapter 3, there are many real biological rhythms — but biorhythms are not part of them.

The notion of biorhythms was developed in the early years of the 20th century by two individuals working independently: German physician Wilhelm Fliess (1859–1928) and Austrian psychologist Hermann Swoboda (1873–1963). Both men were acquainted with Sigmund Freud (1856–1939). Freud, of course, was the creator of psychoanalysis, the man who popularized the notion of a vastly rich unconscious mind. The artistic value of this was clearly expressed by André Breton (1896–1966) in his manifesto launching the surrealist movement in 1924.[1] The scientific value of psychoanalysis, on the other hand, has always been polemical, particularly because of Freud's one-man approach to the development of his theories.[2] The lack of progressive, independent verification by other investigators, which led to the presentation of a whole finished system as a packet fallen from the sky, is what differentiated psychoanalysis from serious scientific research being conducted at the turn of the century. The same was true about the work of Fliess and Swoboda.

According to followers of Fliess and Swoboda, biorhythms are three natural cycles within the human body that affect us physically, emotionally, and intellectually. The three biorhythms begin when a person is born, and they oscillate with absolute precision, as perfect sine waves, until the person dies. The *physical* rhythm regulates physical strength, energy, endurance, sex drive, confidence, and so forth. The *emotional* rhythm governs creativity, sensitivity, mood, and so on. The *intellectual* rhythm is associated with intelligence, memory, mental alertness, logical thinking, etc. The physical rhythm is 23 days long; the emotional, 28 days long; and the intellectual, 33 days long. The different length of the three cycles causes them to be constantly out of phase (they coincide only at birth and every 58 years plus 66 or 67 days thereafter, depending on the number of leap years in between). Thus, a person's disposition on any given day will be a composite of the states of the three rhythms. For each rhythm, the bad moments are not those below the baseline but those at the crossing of the baseline (which are, accordingly, called *critical days*). By calculating and studying one's biorhythms, one is supposedly capable of knowing what to expect each day and, therefore, one is also capable of avoiding bad experiences. Believers are cautioned to watch for critical days — and to avoid doing anything important or risky on *double* critical days. Several self-help books have been published to guide potential followers.[3–12]

To facilitate the discussion, I will concentrate on a specific book, Bernard Gittelson's *Biorhythm: A Personal Science*.[12] This book, first published in 1975, has gone through 10 editions and has sold over one million copies. Gittelson certainly has no intent to deceive the reader. Although the publisher's blurb asserts that biorhythm "is an established scientific discipline," the author himself warns in the preface that biorhythm is controversial and its legitimacy is difficult to prove. According to him, the book's subtitle (*A Personal Science*) is intended to convey the idea that the reader is to be the "scientist" and to judge whether biorhythm is worth anything. Later on, he is honest enough to acknowledge that "when consulted about biorhythm, three of the most distinguished students of circadian and other biological rhythms have been quick to dismiss the theory and to distinguish it from their own work." Yet, he must make substantial claims if he expects people to buy the book. And he does:

> "If only" … There are few more common or poignant phrases. We speak those words when we have accidents, miss opportunities, witness disasters, or whenever we are regretfully surprised by events in our own lives or in those of others. We would speak them much less often if we had a deeper understanding of the roots of human behavior and the causes of human frailty; some way to increase our knowledge of what people are likely to do and what may happen to

them. It is exactly this kind of knowledge and understanding that biorhythm offers."

Does biorhythm really offer this kind of knowledge and understanding? Obviously, the only way to tell is to conduct tests. Even the most absurd theory can, in principle, be correct. As mentioned in the preceding chapter, a scientific fact is a hypothesis that has been tested repeatedly and has not been refuted. My discussion of real biological rhythms in this book is backed by more than 800 bibliographic references — which is actually only a small representative sample of the scientific literature on the subject. The consideration that biorhythm has not been properly tested after almost 100 years strongly suggests that its proponents are afraid of the outcome. In the absence of proper tests, Gittelson tries to convince the reader by using anecdotal evidence — and he does tell a few good stories. He might just as well have used tales of Greek warriors winning battles after making offerings at Aries' temple as proof of the existence of Olympic gods. Superstition thrives on the reporting of single events and anomalies.

Again, the real test of the scientificity of a theory is empirical research. Nevertheless, there are some clear signs of why biorhythm is unlikely to be more than a farce. The first one, of course, is the fact that serious researchers consider it a farce. But this is very circumstantial evidence. A real problem is biorhythm's gross disregard for the issue of biological variability.

There certainly is a lot of regularity in biological processes. The notion of *homeostasis* — that is, that bodily functions are regulated at a constant level despite alterations in the environment—was developed in the 19th century by the renowned physiologist Claude Bernard,[13] and is such a fundamental physiological principle that its understanding is considered a requisite of cultural literacy.[14] However, the stability of physiological functions implied by the concept of homeostasis is a dynamic process, not a static condition.[15] Even something as mundane as one's bedtime expresses regularity with variability. You probably go to bed at roughly the same time each night (say, 11 o'clock or midnight), but rarely do you keep the accuracy of your bedtime to within minutes (and certainly not seconds). Variability is an essential feature of biological processes. As a matter of fact, absence of variability is often a sign of disease.[16]

As we will see throughout this book, real biological rhythms do have a pattern that allows us to identify them as actual rhythms, but they are clearly subject to biological variability. In contrast, biorhythms are amazingly "clean" rhythms that repeat themselves for the whole life of the individual without ever deviating, even slightly, from a perfect sine wave. And, of course, the use of "biorhythm charts" (which take more than half of Gittelson's 430-page book) is at all possible only because the cycles of 23, 28, and 33 days proceed without a single glitch, with an accuracy better than a tenth of a second, throughout a person's life. This extreme regularity is strongly suggestive of a theory developed in someone's head without any observation of actual biological processes.

Another major sign of biorhythm's flimsiness is its presentation as a panacea — a cure for all ills. For this, I will make use of another biorhythm bestseller, Jacyntha Crawley's *The Biorhythm Book*.[11] Although the list is too long to be repeated here, these are some examples of what Crawley says biorhythms can do for you: determine compatibility with a romantic partner, determine compatibility with co-workers, improve family relationships, predict the day of birth of a child, predict the sex of an unborn child, predict one's day of death, optimize one's creative efforts, improve one's dieting efforts, improve performance in a driving test, choose the right time for immunization shots, reduce the risk of a heart attack or stroke, avoid drug addiction, avoid car accidents, and much more! Those who live in the real world, rather than in a world of fantasy, are well aware that no tool has so many applications. To make such claims is to betray one's intention to give false hopes to naive readers. In a free society, people should certainly have the right to publish whatever they want, but free readers should exercise their right to ignore the material.

Once again, absurd as biorhythm is, it could conceivably be true. This is certainly the reason that millions of people buy these books. Until disproved, any statement of fact — no matter how absurd — could be true. Indeed, the National Institutes of Health have recently upgraded its six-year-old office of alternative medicine to a full-fledged center (the National Center for Complementary and Alternative Medicine) destined to evaluate the potential medical use of procedures that seem too good to be true.[17] After all, they *may* actually work. Although the use of any untested medical procedure has its risks,[18,19] it is a fact that many patients — with or without their physician's endorsement — use alternative therapies,[20] and more than half of American medical schools now offer elective courses on alternative medicine.[21] Treatments that are medically ineffective but are not harmful can certainly provide patients with the hope and relief of distress that comes from being in a healing situation.[22] In terms of basic research, except for the diversion of funds from research on more promising medical procedures, there is certainly no harm in testing hypotheses that seem absurd. Perhaps biorhythm supporters might jump on the bandwagon and finally start to put the theory to a real test. As a matter of fact, I am aware of at least one recent study published in the medical literature that attempted to test the biorhythm theory.[23] If more studies are conducted, biorhythm may soon become not only a *refutable* hypothesis but a *refuted* theory as well.

pregnancy predestination

When a single woman — or a married woman who has not had children — enters her third or fourth decade of life and starts having the urge to become a mother, it is not uncommon to say that "her biological clock is ticking." The meaning of the expression is that there is some sort of clock in the woman's body that starts ticking louder and louder as menopause approaches. Supposedly, the woman's life will not be complete unless she becomes pregnant and has a child. Several books about late motherhood have the words "biological clock" in their titles.[24-26] Is this just a metaphor, or is there really some sort of pregnancy predestination?

In a world where every life form has a limited lifespan, it is evident that without reproduction life would soon vanish. In this very general sense, it would be appropriate to speak of pregnancy predestination. If no women had children, and scientists did not develop human cloning right away, human life would soon vanish. But then, natural selection took care of making sex such a pleasurable activity that, for hundreds of thousands of years, humans did not have to fret about the necessity of reproduction — it just came along as a natural consequence of having sex. Therefore, to the extent that one has a "predestination" to enjoy sex, one could speak of pregnancy predestination. And, of course, anyone who has experienced the strong emotions of parenthood can testify that missing on parenthood would be missing on one of the most significant parts of life. But, metaphors aside, is there a biological clock that sets the time for reproduction? Yes and no. The "no" refers to a clock that would tell women to have children early in life. At least, no evidence was ever collected that such a clock might exist. The "yes" refers to a seasonal cycle of reproduction — which is much stronger in other animals than in humans — and to the menstrual cycle. These are real biological rhythms. They will be discussed briefly in the following chapter.

So, if there is no biological clock that tells women to have children early in life, is it OK for mature women to have children? There is no straightforward answer to this question. Although there is no mechanism responsible for the urge to reproduce early in life, people do get older as time goes by, and aging is known to impair various physiological processes. So, the question is whether older women should have children. Evolutionarily, there is little that can be said about it. As recently as the 1850s, the life expectancy at birth of American women was only 40 years (and has risen to 60 years in the 1920s and to 80 years in the 1990s).[27] Consequently, a woman's ability to have children in her forties was not subject to natural selection for practically all of human evolution. The ability of older women to attract suitable partners apparently was hindered by selective forces that led to the evolution of a male preference for younger females,[28] but this is another story.

Challenges associated with late motherhood include fertility problems, high risk of chromosomal defects, high rates of miscarriage, and labor complications. On top of this, of course, there are physical and social tensions. Is it all worth it? This is a decision that the future mother must make — together with the father and the physician. While postponing motherhood to the late thirties and forties is not advisable, the idea becomes much more attractive once one has reached that age without having experienced the joys of motherhood. In her book on late motherhood, Pamela Armstrong quotes a childbirth educator whose opinion is, although purely anecdotal, very compelling:[25]

What comes to my mind when I think of the women I have
worked with who are older than the average is the word
'precious.' All babies are precious. But the pregnancies and
babies for these women are absolutely precious. They were
definitely glowing during their pregnancies. They looked
wonderful.

aging

Generation after generation, people are born, get old, and die. We could
certainly refer to one's life cycle as a biological rhythm. Personally, I do not
think that the usage is appropriate. The standard use of the term *biological
rhythm* is to refer to a cyclic process in a biological system. Usually, what we
mean by "biological system" is an organism (say, a tree, a mouse, or a human
organism). For each organism, of course, life is not a cyclic process. Fantasies
about reincarnation aside, we all are born and die only once. So, although
we may go through a life cycle, this cycle does not repeat itself — and a
cycle that occurs only once does not qualify as a rhythmic process. Thus, if
you go to a given restaurant only once, you cannot say that you go there
regularly. And a clock that stops running after one day is not really a clock
— it is a defective instrument that you will return to the store for a refund.

Of course, we could consider the biological system to be the species
rather than the individual. Species only live once too, but — as long as the
species is not extinct — one generation is followed by the next. Thus, there
is a cycle of life within the species — and the cycle certainly involves aging.
However, the aging of each individual happens only once. No one rejuve-
nates and starts the aging process again. Therefore, aging is not a repetitive
phenomenon and should not be called a biological rhythm.

Even though we age and die only once and, therefore, aging is not a
biological rhythm, the process of aging does involve biological rhythms.
Various studies have shown that aging is accompanied by a disruption in
biological rhythms in laboratory animals[29-34] as well as in humans.[35-40] More
significantly, aging itself is the result of a disruption in the rhythm of cell
reproduction: cells are genetically programmed to undergo a finite number
of replications and, when the limit is reached, hundreds of biological changes
that characterize aging take place.[41]

Now, having talked about what the biological clock and biological
rhythms are *not*, it is about time to talk about what they *are*. Chapter 3 will
briefly describe a variety of biological rhythms and will introduce a particular type of rhythm that is controlled by a mechanism that can be genuinely
called the *biological clock*.

bibliographic notes

1. Breton, A. (1962). *Manifestes du Surrealism*. Paris: J. J. Pauvert.

2. Hergenhahn, B. R. (1997). *An Introduction to the History of Psychology* (3rd edition). Pacific Grove, CA: Brooks/Cole.

3. Wernli, H. J. (1960). *Biorhythm: A Scientific Exploration into the Life Cycles of the Individual.* New York: Crown.

4. Smith, R. E. (1976). *The Complete Book of Biorhythm Life Cycles.* New York: Aardvark.

5. Thomson, H. M. (1976). *Biorhythm for Life.* North Andover, MA: Evergreen.

6. Rebsch, D. (1977). *Biorhythm and You.* Rockville, MD: Universal Biorhythm.

7. Tatai, K. (1977). *Biorhythm for Health Design.* New York: Japan Publications Trading.

8. Gittelson, B. (1977). *Biorhythm Sports Forecasting.* New York: Arco.

9. Bartel, P. C. (1978). *Biorhythm.* New York: F. Watts.

10. Shap, M. (1978). *The Biorhythm Decision Maker and Guide.* New York: Grosset & Dunlap.

11. Crawley, J. (1996). *The Biorhythm Book.* Boston: Journey.

12. Gittelson, B. (1996). *Biorhythm: A Personal Science.* New York: Warner.

13. Bernard, C. (1872). *De la Physiologie Générale.* Paris: Hachette.

14. Hirsch, E. D. (1988). *Cultural Literacy.* New York: Vintage.

15. Mrosovsky, N. (1990). *Rheostasis: The Physiology of Change.* Oxford, UK: Oxford University Press.

16. Goldberger, A. L., Rigney, D. R., and West, B. J. (1990). Chaos and fractals in human physiology. *Scientific American* 262: 42-49.

17. Couzin, J. (1998). Beefed-up NIH center probes unconventional therapies. *Science* 282: 2175-2176.

18. Angell, M. and Kassirer, J. P. (1998). Alternative medicine: The risks of untested and unregulated remedies. *New England Journal of Medicine* 339: 839-841.

19. Borins, M. (1998). The dangers of using herbs: What your patients need to know. *Postgraduate Medicine* 104: 99-100.

20. Wainapel, S. F., Thomas, A. D., and Kahan, B. S. (1998). Use of alternative therapies by rehabilitation outpatients. *Archives of Physical Medicine and Rehabilitation* 79: 1003-1005.

21. Wetzel, M. S., Eisenberg, D. M., and Kaptchuk, T. J. (1998). Courses involving complementary and alternative medicine at U.S. medical schools. *Journal of the American Medical Association* 280: 784-787.

22. Brown, W. A. (1998). Alternative medicine: It's time to get smart. *Scientist* 12(24): 13.

23. D'Andrea, V. J., Black, D. R., and Stayrook, N. G. (1984). Relation of the Fliess-Swoboda biorhythm theory to suicide occurrence. *Journal of Nervous and Mental Diseases* 172: 490-494.

24. McKaughan, M. (1987). *The Biological Clock: Reconciling Careers and Motherhood in the 1980's.* New York: Doubleday.

25. Armstrong, P. (1996). *Beating the Biological Clock: The Joys and Challenges of Late Motherhood.* London, UK: Headline.

26. Paulson, R. J. and Sachs, J. (1998). *Rewinding Your Biological Clock: Motherhood Late in Life.* San Francisco: W. H. Freeman.

27. Brunner, B. (1998). *Time Almanac.* Boston: Information Please.

28. Buss, D. M. (1989). Sex differences in human mate preferences: Evolutionary hypotheses tested in 37 cultures. *Behavioral and Brain Sciences* 12: 1-14.

29. Halberg, J., Halberg, E., Regal, P., and Halberg, F. (1981). Changes with age characterize circadian rhythm in telemetered core temperature of stroke-prone rats. *Journal of Gerontology* 36: 28-30.
30. Morin, L. P. (1988). Age-related changes in hamster circadian period, entrainment, and rhythm splitting. *Journal of Biological Rhythms* 3: 237-248.
31. Refinetti, R., Ma, H., and Satinoff, E. (1990). Body temperature rhythms, cold tolerance, and fever in young and old rats of both genders. *Experimental Gerontology* 25: 533-543.
32. Witting, W., Mirmiran, M., Bos, N. P. A., and Swaab, D. F. (1994). The effect of old age on the free-running period of circadian rhythms in rats. *Chronobiology International* 11: 103-112.
33. Li. H. and Satinoff, E. (1995). Changes in circadian rhythms of body temperature and sleep in old rats. *American Journal of Physiology* 269: R208-R214.
34. Hurd, M. W. and Ralph, M. R. (1998). The significance of circadian organization for longevity in the golden hamster. *Journal of Biological Rhythms* 13: 430-436.
35. Lobban, M. C. and Tredre, B. E. (1967). Diurnal rhythms of renal excretion and of body temperature in aged subjects. *Journal of Physiology* 188: 48P-49P.
36. Weitman, E. D., Moline, M. L., Czeisler, C. A., and Zimmerman, J. C. (1982). Chronobiology of aging: Temperature, sleep-wake rhythms and entrainment. *Neurobiology of Aging* 3: 299-309.
37. Touitou, Y., Reinberg, A., Bogdan, A., Auzéby, A., Beck, H., and Touitou, C. (1986). Age-related changes in both circadian and seasonal rhythms of rectal temperature with special reference to senile dementia of Alzheimer type. *Gerontology* 32: 110-118.
38. Vitiello, M. V., Smallwood, R. G., Avery, D. H., Pascualy, R. A., Martin, D. C., and Prinz, P. N. (1986). Circadian temperature rhythms in young and aged men. *Neurobiology of Aging* 7: 97-100.
39. Campbell, S. S., Gillin, J. C., Kripke, D. F., Erikson, P., and Clopton, P. (1989). Gender differences in the circadian temperature rhythms of healthy elderly subjects: Relationships to sleep quality. *Sleep* 12: 529-536.
40. Nakazawa, Y., Nonaka, K., Nishida, N., Hayashida, N., Miyahara, Y., Kotorii, T., and Matsuoka, K. (1991). Comparison of body temperature rhythms between healthy elderly and healthy young adults. *Japanese Journal of Psychiatry and Neurology* 45: 37-43.
41. Hayflick, L. (1989). Antecedents of cell aging research. *Experimental Gerontology* 24: 355-365.

chapter three

---

biological rhythms

> In this chapter, the diversity of biological rhythms (with cycles ranging from a few seconds to a whole year) is described. The great importance of one type of biological rhythm — the circadian rhythm — is emphasized and the notion of the biological clock is introduced.

Before ever opening this book, you may have observed — or at least heard about — a variety of biological rhythms. Some of the rhythms you have observed or heard about are probably real rhythms (such as your heartbeat), whereas others are pure superstition (such as biorhythm). This chapter will describe many rhythms. Why should you trust that the rhythms I talk about are real rhythms? Primarily because I will cite several scientific studies in support of my statements. However, I cannot possibly provide enough supporting evidence in a short chapter. Therefore, I do not expect you to be fully convinced until you read the following seven chapters, where I provide enough evidence to convince even the most skeptical of readers. For the time being, your skeptical attitude is totally acceptable.

Biological rhythms are "recurrent events within a biological system."[1] By *recurrent events* we mean biological processes that take place more than once — usually many times — throughout the system's life. By *biological system* we usually mean an organism (an individual). However, the expression can be applied to groups of individuals, as when one considers the seasonal rhythm of spouse abuse, which peaks in the summer,[2] or the weekly rhythm of myocardial infarctions (heart attacks).[3] The frequency of recurrence of the events may be very high (once every fraction of a second) or very low (once each year or once every several years). Three examples of biological rhythms are illustrated in Figure 1. The electrical activity of many neurons (nerve cells) in the central nervous system exhibits a constant rhythm of firing rate (action potentials) in the absence of perturbations. The particular neuron whose activity is depicted in Figure 1 fires once approximately every 50

NEURON

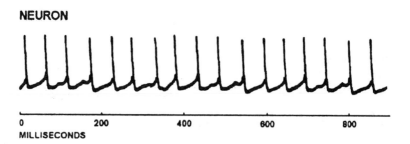

```
0          200         400         600         800
MILLISECONDS
```

HEART

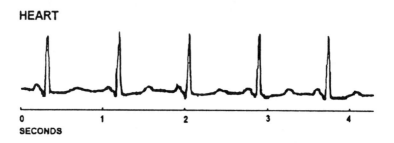

```
0          1           2           3           4
SECONDS
```

BODY

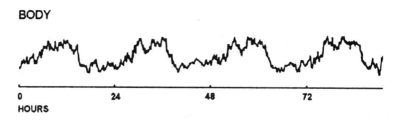

```
0          24          48          72
HOURS
```

Figure 1. Examples of biological rhythms. *Neuron* refers to the spontaneous electrical activity of a typical neuron in the central nervous system. *Heart* refers to the electro-cardiogram of an adult human at rest. *Body* refers to the daily rhythmicity found in numerous processes in the body, such as body temperature.

milliseconds (frequency of 20 Hz). The electrical activity of the human heart is also rhythmic, as indicated by the electrocardiogram reproduced in the figure. In a resting subject, the activity pattern repeats itself a little more often than once a second (approximately 70 beats per minute). The frequency of heartbeat is actually inversely related to body size, so that in very small animals resting heart rates can be as high as 600 beats per minute (10 Hz).[4] The third rhythm shown in Figure 1 is generically named "body" because it can be observed in numerous organs throughout the body. The specific variable represented in Figure 1 is the body temperature of a laboratory rat. The oscillation repeats itself approximately once every 24 hours (once a day).

Biological processes that repeat themselves every 24 hours are called *daily* or *circadian* (sir-kay-de-an) rhythms. As detailed in chapter 4, *circadian* is not exactly the same as *daily*; however, the distinction can be ignored for now. Circadian rhythms are so pervasive in the body that they affect the operation of other rhythms. For instance, Figure 2 illustrates the fact that the rhythm of a heart beating is itself subject to circadian rhythmicity. In other words, there is a circadian rhythm of modulation of heartbeat. Thus, strictly speaking, one cannot say that the heart rate of the golden hamster is 400 beats per minute (bpm), or any other exact value. Although the statement would be true as a daily average, heart rates actually oscillate more or less regularly throughout the day from values as low as 360 bpm to values as high as 480 bpm. Consequently, although the average heart rate of the hamster is 400 bpm, it is much lower during the day than during the night (when the animal is active). Also shown in the figure are the body temperature records for the same animal. Although the mean body temperature of this hamster is approximately 36.7°C (98.1°F), values oscillate daily from as low as 36.3°C to as high as 38.0°C. Human body temperature also oscillates daily (from approximately 36.5 to 37.4°C),[5-9] which means that the traditional definition of fever as a temperature above 37.0°C (98.6°F) applies only to certain times of the day. At about 5 pm, a temperature reading of 37.0°C not only is not a fever but is actually below the expected (normal) reading of 37.4°C.

Biological rhythms with frequencies higher than that of circadian rhythms (such as the electrocardiogram) are called *ultradian* rhythms. Rhythms with frequencies lower than that of circadian rhythms (such as the menstrual cycle) are called *infradian* rhythms. Notice that, since period is the inverse of frequency, the *period* of infradian rhythms is *longer* than that of circadian rhythms (for instance, the menstrual cycle, which lasts approximately 29 days, is longer than the circadian rhythm, which lasts only one day). Of course, *longer* and *shorter* are relative terms. What are the limits of circadian rhythms? Usually, rhythms with periods longer than 30 hours are called infradian, whereas rhythms with periods shorter than 18 hours are called ultradian. However, as you might expect, there are exceptions to this rule — and some of them will be mentioned below.

Besides well-known ultradian rhythms such as heart rate and breathing rate, regular oscillations with periods of a few minutes or hours have been reported many times in the scientific literature. For instance, small oscillations in human skin temperature with a period of a few minutes to half an hour have been reported.[10] Also reported has been a three-and-a-half-hour activity rhythm in voles.[11] An ultradian rhythm of carbon dioxide production with a period of about an hour has been reported in the mouse, quail, rat, guinea pig, chicken, and monkey.[12] Oscillations with a period of about an

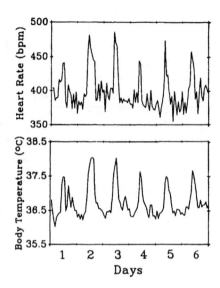

Figure 2. Daily rhythms of heart rate and body temperature of a golden hamster. Measurements were taken hourly (by telemetry) for six consecutive days.The animal had free access to a running wheel.

hour were also reported for body temperature and blood levels of insulin and urea in dairy cows.[13] A 12-hour ultradian component of the body temperature rhythm has been reported in humans.[14] Several studies in rodents have identified ultradian components of the locomotor activity and body temperature rhythms that are submultiples of the circadian period (such as 12, 8, 6, and 4.8 hours),[15–18] Such ultradian oscillations seem to be stronger (greater amplitude) during the night (when the animals are active) than during the day.[19]

Humans often use watches and clocks to measure the passage of time. These instruments allow us to time events without having to constantly think about them. However, we are quite capable of estimating the passage of time in the range of seconds, minutes, or hours.[20] Interestingly, this capability seems to depend on our circadian rhythms. Thus, estimates of the duration of an hour are longer when the person lives under artificial conditions where a day is longer than 24 hours.[21] In other words, we perceive an hour as 1/24 of a day, not necessarily as 60 minutes. Estimates of brief intervals (a few seconds) are not affected the same way.[22]

Infradian rhythms are as varied as ultradian rhythms. Weekly and half-weekly rhythms have been described in human subjects.[3] In animals that live close to the ocean and graze in intertidal areas, tidal rhythms (period of 12.4 hours) and lunar rhythms (period of 24.8 hours) have been observed.[23] Especially well studied have been the biological rhythms associated with reproduction, which include not only the ovarian cycle but also seasonal cycles (*circannual* rhythms). The duration of the ovarian cycle varies greatly from one species to another. While the human *menstrual* cycle lasts approximately 29 days, the ovarian cycle lasts 1 day in the chicken,[24] 4 days in the golden hamster,[25] 31 days in the orangutan,[26] and 45 days in the killer whale.[27] Often, the ovarian cycle has multiple effects; for example, in the golden hamster, 4-day *estrous* periodicity has been demonstrated not only in the modulation of blood hormone levels and occurrence of ovulation,[25,28] but also in the pattern of vaginal secretion,[29,30] in the initiation of behavioral sexual receptivity to males,[31–32] and in the amount and temporal organization of running-wheel activity.[33, 34] In humans, sexual receptivity is not tied to the ovarian cycle, although a woman's sexual appetite does seem to increase at the time of ovulation.[35] The menstrual cycle is strongly associated with small alterations in body temperature, and the rise in basal temperature immediately following ovulation has long been used for contraception or its very opposite, the treatment of infertility.[36–39]

Seasonal cycles affect reproduction in many species. For instance, in hamsters and voles, the gonads undergo regression during the winter, so that the animals are physically unable to reproduce. One might expect this to be due to the lower environmental temperature, but it is actually due to the shortening of daylength (photoperiod).[40,41] The same is true for tiny crustaceans such as the pill bug[42] as well as for large ungulates such as the deer.[43] Although similar changes do not occur in humans, there is a small but measurable seasonal rhythm in human births.[44]

Many physiological functions in addition to reproduction undergo seasonal changes. These include body weight gain or loss, pelage molt, alterations in the level of locomotor activity, and, of course, drastic changes in body temperature in hibernators.

Several species of birds and small mammals enter daily torpor, especially if food-deprived, regardless of season.[45–52] In the mouse lemur (which is the smallest primate, weighing as much as a mouse), the daily fall in body temperature was estimated to provide energy savings of about 60%.[53] Of course, energy conservation is what hibernation is all about. With very low ambient temperatures and scarce food during the winter, many animals depend on hibernation to survive through the spring. Hibernators include Syrian (golden) hamsters,[54] Turkish hamsters,[55] European hamsters,[56] bats,[57] chipmunks,[58] hedgehogs,[59,60] ground squirrels,[61,62] and many others. Bears experience prolonged sleep periods in the winter but do not exhibit the drastic reduction in metabolic rate and body temperature of true hibernators.[63, 64] Hibernation is a regulated hypometabolic state, which means that the control of body temperature is not simply turned off during hibernation; rather, temperature is regulated at a much lower level.[65] The circadian rhythm of body temperature persists during hibernation, although, of course, at a much lower mean level.[66–68]

Some seasonal rhythms may seem rather obvious given the harshness of winter in many regions. For instance, it is not surprising to learn that beavers remain mostly inactive inside their lodges during the Canadian winter, even though they do not hibernate.[69] However, in several instances, it has been shown that the seasonal changes in biological functions are not due to variations either in temperature or in daylength, even though these variations may have a fine-tuning effect on the biological functions. Fantastic as it may sound, there is a biological clock that generates circannual rhythms. Supposedly, this clock evolved as a mechanism to anticipate the seasonal changes in the environment, but it now functions on its own. For instance, reproductive status and molting of the stonechat (a bird) were monitored consecutively for more than 7 years in the laboratory under conditions of constant daylength and constant environmental temperature. Gonad size and molt exhibited seasonal rhythmicity despite the constancy of the environment.[70] Of course, the experimenters could have overlooked some environmental variable that provided the birds with temporal cues. However, the period of the circannual rhythms exhibited by the birds was not 12 months (as one would expect if geophysical phenomena were providing a time cue) but 9 months. Thus, the birds must have had an internal clock that, in the absence of external cues to finetune it, ran with a period of 9 months. Does it sound implausible? Perhaps. However, a similar type of circannual rhythm was measured in other species in other laboratories, such as chipmunks (period = 11 months),[71] ground squirrels (period = 10 to 11 months),[72,73] and ewes (period = 14 months).[74] If you think about it, a circannual clock is no more implausible than an ultradian clock — and I assume that you have no problem accepting the idea that your heart has a clock. It has been known for many decades that the periodicity of heart beating is driven by a pacemaker located in the sinus node of the heart and modulated by the sympathetic and parasympathetic nervous systems.[75-78]

In addition to the cardiac pacemaker and the circannual clock, a pacemaker located in the brainstem is responsible for the breathing rhythm.[79] The existence of the ovarian cycle implies the existence of yet another pacemaker, even though this pacemaker is often strongly modulated by the circadian pacemaker.[80, 81] Oh, did I forget to mention it? There is also a *circadian* pacemaker (to be discussed in the next seven chapters). Watch out: there are clocks everywhere.

Now you must be confused. Up to this point, I have been using the expression *the biological clock* freely. Now I am saying that there are many biological clocks. The truth is that there are indeed many biological clocks. However, I believe it is very appropriate to refer to the *circadian* clock as *the* clock for several reasons:

1. Circadian rhythms are robust rhythms that can be observed without the necessity of subtle manipulations to extract them out of biological noise.
2. Circadian rhythmicity is a strong determinant of our life habits, such as getting up and going to sleep each day.
3. Circadian rhythms affect us literally every day (unlike the menstrual cycle, which is practically ignored for most days of each month).

4. The circadian clock affects the operation of practically all functions in the body (unlike the cardiac pacemaker, for instance, which affects only heartbeat).
5. The properties of the circadian clock are well known, and its physical location (in mammalian species) has been anatomically identified (unlike the circannual pacemaker, whose location has yet to be determined).

Of course, there is a sixth reason: circadian rhythms are my professional passion — I love to study them. But you do not need to share my passion in order to appreciate the beauty and the importance of circadian rhythms. The remaining chapters of this book will tell you all about circadian rhythms — what they are, how they are organized, how they are controlled, and how knowing about them can help us improve our lives. Although the current incompleteness of scientific knowledge of biological rhythms may sometimes be disappointing — especially if compared with the unrealistic claims made in the popular literature — the understanding of real facts will provide you with the basis for a justified optimism about the multiple applications of fundamental knowledge about the biological clock.

bibliographic notes

1. Aschoff, J. (1981). A survey of biological rhythms. In: Aschoff, J. (Ed.). *Handbook of Behavioral Neurobiology, Volume 4: Biological Rhythms.* New York: Plenum, pp. 3-10.
2. Michael, R. P. and Zumpe, D. (1986). An annual rhythm in the battering of women. *American Journal of Psychiatry* 143: 637-640.
3. Cornélissen, G., Breus, T. K., Bingham, C., Zaslavskaya, R., Varshitsky, M., Mirsky, B., Teibloom, M., Tarquini, B., Bakken, E., and Halberg, F. (1993). Beyond circadian chronorisk: Worldwide circaseptan-circasemiseptan patterns of myocardial infarctions, other vascular events, and emergencies. *Chronobiologia* 20: 87-115.
4. Stahl, W. R. (1967). Scaling of respiratory variables in mammals. *Journal of Applied Physiology* 22: 453-460.
5. Davy, J. (1845). On the temperature of man. *Philosophical Transactions of the Royal Society of London* 135: 319-333.
6. Ogle, W. (1866). On the diurnal variations in the temperature of the human body in health. *St. George's Hospital Reports* 1: 221-245.
7. Mellette, H. C., Hutt, B. K., Askovitz, S. I., and Horvath, S. M. (1951). Diurnal variations in body temperature. *Journal of Applied Physiology* 3: 665-675.
8. Aschoff, J., Gerecke, U., and Wever, R. (1967). Phasenbeziehungen zwischen den circadianen Perioden der Aktivität und der Kerntemperatur beim Menschen. *Pflügers Archiv* 295: 173-183.
9. Scales, W. E., Vander, A. J., Brown, M. B., and Kluger, M. J. (1988). Human circadian rhythms in temperature, trace metals, and blood variables. *Journal of Applied Physiology* 65: 1840-1846.
10. Doust, J. W. L. (1979). Periodic homeostatic fluctuations of skin temperature in the sleeping and waking state. *Neuropsychobiology* 5: 340-347.

circadian physiology

11. Halle, S. (1995). Diel pattern of locomotor activity in populations of root voles, Microtus oeconomus. *Journal of Biological Rhythms* 10: 211-224.
12. Stupfel, M., Gourlet, V., Perramon, A., Mérat, P., Putet, G., and Court, L. (1995). Comparison of ultradian and circadian oscillations of carbon dioxide production by various endotherms. *American Journal of Physiology* 268: R253-R265.
13. Lefcourt, A. M., Huntington, J. B., Akers, R. M., Wood, D. L., and Bitman, J. (1999). Circadian and ultradian rhythms of body temperature and peripheral concentration of insulin and nitrogen in lactating dairy cows. *Domestic Animal Endocrinology* 16: 41-55.
14. Kronauer, R. E. and Jewett, M. E. (1992). The relationship between circadian and hemicircadian components of human endogenous temperature rhythms. *Journal of Sleep Research* 1: 88-92.
15. Wollnik, F. and Turek, F. W. (1988). Estrous correlated modulations of circadian and ultradian wheel-running activity rhythms in LEW/Ztm rats. *Physiology and Behavior* 43: 389-396.
16. Gerkema, M. P., Groos, G. A., and Daan, S. (1990). Differential elimination of circadian and ultradian rhythmicity by hypothalamic lesions in the common vole, Microtus arvalis. *Journal of Biological Rhythms* 5: 81-95.
17. Ortega, G. J., Romanelli, L., and Golombek, D. A. (1994). Statistical and dynamical analysis of circadian rhythms. *Journal of Theoretical Biology* 169: 15-21.
18. Refinetti, R. (1996). Ultradian rhythms of body temperature and locomotor activity in wild-type and tau-mutant hamsters. *Animal Biology* 5: 111-115.
19. Refinetti, R. (1994). Circadian modulation of ultradian oscillation in the body temperature of the golden hamster. *Journal of Thermal Biology* 19: 269-275.
20. Morell, V. (1996). Setting a biological stopwatch. *Science* 271: 905-906.
21. Aschoff, J. (1992). On the dilatability of subjective time. *Perspectives in Biology and Medicine* 35: 277-280.
22. Aschoff, J. (1998). Human perception of short and long time intervals: Its correlation with body temperature and the duration of wake time. *Journal of Biological Rhythms* 13: 437-442.
23. Wikelski, M. and Hau, M. (1995). Is there an endogenous tidal foraging rhythm in marine iguanas? *Journal of Biological Rhythms* 10: 335-350.
24. Winget, C. M., Averkin, E. G., and Fryer, T. B. (1965). Quantitative measurement by telemtry of ovulation and oviposition in the fowl. *American Journal of Physiology* 209: 853-858.
25. Alleva, J. J., Waleski, M. V., and Alleva, F. R. (1971). A biological clock controlling the estrous cycle of the hamster. *Endocrinology* 88: 1368-1379.
26. Asa, C. S., Fischer, F., Carrasco, E., and Puricelli, C. (1994). Correlation between urinary pregnanediol glucuronide and basal body temperature in female orangutans, Pongo pygmaeus. *American Journal of Primatology* 34: 275-281.
27. Walker, L. A., Cornell, L., Dahl, K. D., Czekala, N. M., Dargen, C. M., Joseph, B., Hsueh, A. J. W., and Lasley, B. L. (1988). Urinary concentrations of ovarian steroid hormone metabolites and bioactive follicle-stimulating hormone in killer whales (Orcinus orchus) during ovarian cycles and pregnancy. *Biology of Reproduction* 39: 1013-1020.
28. Ciaccio, L. A. and Lisk, R. D. (1971). Hormonal control of cyclic estrus in the female hamster. *American Journal of Physiology* 221: 936-942.

29. Orsini, M. W. (1961). The external vaginal phenomena characterizing the stages of the estrous cycle, pregnancy, pseudo-pregnancy, lactation, and the anestrous hamster, *Mesocricetus auratus. Proceedings of the Animal Care Panel* 11: 193-206.

30. Seegal, R. F. and Goldman, B. D. (1975). Effects of photoperiod on cyclicity and serum gonadrotopins in the Syrian hamster. *Biology of Reproduction* 12: 223-231.

31. Fitzgerald, K. M. and Zucker, I. (1976). Circadian organization of the estrous cycle of the golden hamster. *Proceedings of the National Academy of Sciences* 73: 2923-2927.

32. Swann, J. and Turek, F. W. (1982). Cycle of lordosis behavior in female hamsters whose circadian activity rhythm has split into two components. *American Journal of Physiology* 243: R112-R118.

33. Richards, M. P. M. (1966). Activity measured by running wheels and observation during the oestrous cycle, pregnancy and pseudopregnancy in the golden hamster. *Animal Behaviour* 14: 450-458.

34. Takahashi, J. S. and Menaker, M. (1980). Interaction of estradiol and progesterone: Effects on circadian locomotor rhythm of female golden hamsters. *American Journal of Physiology* 239: R497-R504.

35. Adams, D. B., Gold, A. R., and Burt, A. D. (1978). Rise in female-initiated sexual activity at ovulation and its suppression by oral contraceptives. *New England Journal of Medicine* 299: 1145-1150.

36. Rubenstein, B. B. and Lindsley, D. B. (1937). Relation between human vaginal smears and body temperatures. *Proceedings of the Society for Experimental Biology and Medicine* 35: 618-619.

37. Martin, P. L. (1943). Detection of ovulation by the basal temperature curve with correlating endometrial studies. *American Journal of Obstetrics and Gynecology* 46: 53-62.

38. Kleitman, N. and Ramsaroop, A. (1948). Periodicity in body temperature and heart rate. *Endocrinology* 43: 1-20.

39. Martinez, A. R., van Hooff, M. H. A., Schoute, E., van der Meer, M., Broekmans, F. J. M., and Hompes, P. G. A. (1992). The reliability, acceptability and applications of basal body temperature (BBT) records in the diagnosis and treatment of infertility. *European Journal of Obstetrics and Gynecology and Reproductive Biology* 47: 121-127.

40. Gaston, S. and Menaker, M. (1967). Photoperiodic control of hamster testis. *Science* 158: 925-928.

41. Dark, J., Zucker, I., and Wade, G. N. (1983). Photoperiodic regulation of body mass, food intake, and reproduction in meadow voles. *American Journal of Physiology* 245: R334-R338.

42. Madhavan, K. and Shribbs, J. M. (1981). Role of photoperiod and low temperature in the control of ovigerous molt in the terrestrial isopod, *Armadillidium vulgare. Crustaceana* 41: 263-270.

43. Loudon, A. S., Milne, J. A., Curlewis, J. D., and McNeilly, A. S. (1989). A comparison of the seasonal hormone changes and patterns of growth, voluntary food intake and reproduction in juvenile and adult red deer (*Cervus elaphus*) and Père David's deer (*Elephurus davidianus*) hinds. *Journal of Endocrinology* 122: 733-745.

44. Roenneberg, T. and Aschoff, J. (1990). Annual rhythm of human reproduction. I. Biology, sociology, or both? *Journal of Biological Rhythms* 5: 195-216.

45. Phillips, N. H. and Berger, R. J. (1991). Regulation of body temperature, metabolic rate, and sleep in fasting pigeons diurnally infused with glucose or saline. *Journal of Comparative Physiology B* 161: 311-318.

46. Hohtola, E., Hissa, R., Pyörnilä, A., Rintamäki, H., and Saarela, S. (1991). Nocturnal hypothermia in fasting Japanese quail: The effect of ambient temperature. *Physiology and Behavior* 49: 563-567.

47. Prinzinger, R., Schleucher, E., and Preßmar, A. (1992). Langzeittelemetrie der Körpertemperatur mit synchroner Bestimmung des Energiestoffwechsels beim Blaunackenmausvogel (*Urocolius macrourus*) unter Normal- und Lethargiebedingungen (Torpor). *Journal of Ornithology* 133: 446-450.

48. Hudson, J. W. (1965). Temperature regulation and torpidity in the pygmy mouse, *Baiomys taylori*. *Physiological Zoology* 38: 243-254.

49. Nestler, J. R. (1990). Relationships between respiratory quotient and metabolic rate during entry to and arousal from daily torpor in deer mice (*Peromyscus maniculatus*). *Physiological Zoology* 63: 504-515.

50. Heldmaier, G. and Ruf, T. (1992). Body temperature and metabolic rate during natural hypothermia in endotherms. *Journal of Comparative Physiology B* 162: 696-706.

51. Song, X., Körtner, G., and Geiser, F. (1995). Reduction of metabolic rate and thermoregulation during daily torpor. *Journal of Comparative Physiology B* 165: 291-297.

52. Walker, J. M., Garber, A., Berger, R. J., and Heller, H. C. (1979). Sleep and estivation (shallow torpor): Continuous processes of energy conservation. *Science* 204: 1098-1100.

53. Ortmann, S., Heldmaier, G., Schmid, J. and Ganzhorn, J. U. (1997). Spontaneous daily torpor in Malagasy mouse lemurs. *Naturwissenschaften* 84: 28-32.

54. Lyman, C. P. (1954). Activity, food consumption and hoarding in hibernators. *Journal of Mammalogy* 35: 545-552.

55. Lyman, C. P., O'Brien, R. C., Greene, G. C., and Papafrangos, E. D. (1981). Hibernation and longevity in the Turkish hamster *Mesocricetus brandti*. *Science* 212: 668-670.

56. Wollnik, F. and Schmidt, B. (1995). Seasonal and daily rhythms of body temperature in the European hamster (*Cricetus cricetus*) under semi-natural conditions. *Journal of Comparative Physiology B* 165: 171-182.

57. Menaker, M. (1964). Frequency of spontaneous arousal from hibernation in bats. *Nature* 203: 540-541.

58. Pivorun, E. B. (1976). A biotelemetry study of the thermoregulatory patterns of *Tamias striatus* and *Eutamias minimus* during hibernation. *Comparative Biochemistry and Physiology* 53A: 265-271.

59. Fowler, P. A. and Racey, P. A. (1990). Daily and seasonal cycles of body temperature and aspects of heterothermy in the hedgehog *Erinaceus europaeus*. *Journal of Comparative Physiology B* 160: 299-307.

60. Saboureau, M., Vignault, M. P., and Ducamp, J. J. (1991). L'hibernation chez le Hérisson (*Erinaceus europaeus*) dans son environnement naturel: Étude par biotélémétrie des variations de la température corporelle. *Comptes Rendus de l'Academie des Sciences de Paris Série III* 313: 93-100.

61. Twente, J. W. and Twente, J. A. (1965). Effects of core temperature upon duration of hibernation of *Citellus lateralis*. *Journal of Applied Physiology* 20: 411-416.

62. Barnes, B. M. (1989). Freeze avoidance in a mammal: Body temperatures below 0°C in an Arctic hibernator. *Science* 244: 1593-1595.
63. Hock, R. J. (1960). Seasonal variations in physiologic functions of arctic ground squirrels and black bears. *Bulletin of the Museum of Comparative Zoology* 124: 155-171.
64. Nelson, R. A., Wahner, H. W., Jones, J. D., Ellefson, R. D., and Zollman, P. E. (1973). Metabolism of bears before, during, and after winter sleep. *American Journal of Physiology* 224: 491-496.
65. French, A. R. (1988). The patterns of mammalian hibernation. *Scientific American* 76: 568-575.
66. Menaker, M. (1959). Endogenous rhythms of body temperature in hibernating bats. *Nature* 184: 1251-1252.
67. Florant, G. L., Rivera, M. L., Lawrence, A. K., and Tamarkin, L. (1984). Plasma melatonin concentrations in hibernating marmots: Absence of a plasma melatonin rhythm. *American Journal of Physiology* 247: R1062-R1066.
68. Grahn, D. A., Miller, J. D., Houng, V. S., and Heller, H. C. (1994). Persistence of circadian rhythmicity in hibernating ground squirrels. *American Journal of Physiology* 266: R1251-R1258.
69. Dyck, A. P. and MacArthur, R. A. (1992). Seasonal patterns of body temperature and activity in free-ranging beaver (*Castor canadensis*). *Canadian Journal of Zoology* 70: 1668-1672.
70. Gwinner, E. and Dittami, J. (1990). Endogenous reproductive rhythms in a tropical bird. *Science* 249: 906-908.
71. Richter, C. P. (1978). Evidence for existence of a yearly clock in surgically and self-blinded chipmunks. *Proceedings of the National Academy of Sciences* 75: 3517-3521.
72. Ruby, N. F., Dark, J., Heller, H. C., and Zucker, I. (1996). Ablation of suprachiasmatic nucleus alters timing of hibernation in ground squirrels. *Proceedings of the National Academy of Sciences*: 93: 9864-9868.
73. Lee, T. M. and Zucker, I. (1991). Suprachiasmatic nucleus and photic entrainment of circannual rhythms in ground squirrels. *Journal of Biological Rhythms* 6: 315-330.
74. Thrun, L. A., Moenter, S. M., O'Callaghan, D., Woodfill, C. J. I., and Karsch, F. J. (1995). Circannual alterations in the circadian rhythm of melatonin secretion. *Journal of Biological Rhythms* 10: 42-54.
75. Meek, W. J. and Eyster, J. A. E. (1914). Experiments on the origin and propagation of the impulse in the heart. IV. The effect of vagal stimulation and of cooling on the location of the pacemaker within the sino-auricular node. *American Journal of Physiology* 34: 368-383.
76. Bozler, E. (1943). The initiation of impulses in cardiac muscle. *American Journal of Physiology* 138: 273-282.
77. Hutter, O. F. and Trautwein, W. (1956). Vagal and sympatetic effects on the pacemaker fibers in the sinus venosus of the heart. *Journal of General Physiology* 39: 715-733.
78. Anumonwo, J. M., Delmar, M., Vinet, A., Michaels, D. C., and Jalife, J. (1991). Phase resetting and entrainment of pacemaker activity in single sinus nodal cells. *Circulation Research* 68: 1138-1153.
79. Smith, J. C., Ellenberger, H. H., Ballanyi, K., Richter, D. W., and Feldman, J. L. (1991). Pre-Bötzinger complex: A brainstem region that may generate respiratory rhythm in mammals. *Science* 254: 726-729.

80. Refinetti, R. and Menaker, M. (1992). Evidence for separate control of circadian and estrous periodicity in the golden hamster. *Behavioral and Neural Biology* 58: 27-36.

81. Härter, L. and Erkert, H. G. (1993). Alteration of circadian period length does not influence the ovarian cycle length in common marmosets, *Callithrix j. jacchus* (Primates). *Chronobiology International* 10: 165-175.

part II

---

*circadian rhythms:
physiology and behavior*

chapter four

daily and circadian rhythms

The existence of daily rhythms in a multitude of biological functions in a variety of animal species is documented. The distinction between *daily* and *circadian*
rhythms is made, and the concept of the *circadian pacemaker* is explored.

A surprisingly large number of medical conditions display daily rhythmicity: infant mortality immediately after birth is higher in the late afternoon
and early evening than during the rest of the day;[1] adult suicides are more
frequent in the late morning and early afternoon;[2] injuries caused by firearms
and knives are produced more often around midnight;[3] calls for helicopter
ambulance transportation are more common during daylight hours;[4] unexpected sudden deaths after abdominal surgery are more frequent between
midnight and sunrise[5] ...

Aside from the satisfaction of a morbid curiosity, do these epidemiological findings provide us with useful information? Well, they certainly call
our attention to the fact that there is more in a day than just the alternation
of sunlight and darkness. It is true that your house does not move from one
part of town to another and back during the course of a day, and the number
of molecules in one mole of nitrogen does not change from daytime to
nighttime; but many things oscillate daily. If you are a typical human being,
you most likely walk around and do things during the day and sleep at
night — you are a *diurnal* organism. If you have mice in your attic, you
probably know that they are active during the night and rest during the day
— they are *nocturnal* organisms. Regular, daily oscillations in activity have
been recorded under controlled conditions in a variety of species, including
fruitflies,[6] marine snails,[7] crawfish ,[8] goldfish,[9] lizards,[10] chicken,[11] pigeons,[12]
finches,[13] starlings,[14] mice,[15] rats,[16] mole rats,[17] hamsters,[18] guinea pigs,[19]
squirrels,[20] rabbits,[21] beavers,[22] marmosets,[23] and rhesus monkeys.[24]

Not only the level of activity oscillates daily. In various species, numerous physiological variables were found to display daily rhythms, such as
eating and drinking,[25-28] metabolic rate,[29-31] heart rate and blood pressure,[32-34]

urinary excretion,[20,26,35] hormone concentration in the blood,[36-38] tissue mitoses,[39,40] bioluminescence,[41,42] ambient temperature selection,[43-45] and many others. Particularly well studied has been the rhythm of body temperature, probably because it is very robust and stable in mammals and birds and is relatively easy to measure. Several studies of the daily rhythm of body temperature were conducted as early as the mid-1800s.[46-48] Since then, rhythms of body temperature have been recorded in chicken,[49] pigeons,[50] bats,[20] tenrecs,[51] tree shrews,[52] laboratory mice,[53] pouched mice,[54] spiny mice,[55] Australian antechinuses,[56] Tasmanian bettongs,[57] laboratory rats,[58] tree rats,[59] hamsters,[60] Chilean degus,[61] gerbils,[62] raccoons,[20] rabbits,[63] woodchucks,[64] beavers,[65] cats,[66] dogs,[67] sheep,[68] buffalos,[69] camels,[69] rhesus monkeys,[70] squirrel monkeys,[71] and many other species.

As can be seen in Figure 3, the pattern of the daily oscillation of body temperature is different in different species. While disregarding the high-frequency, irregular oscillations (which may reflect ultradian rhythms or mere biological noise), one can notice robust daily rhythms in the body temperature of all three species. In nocturnal animals (such as the laboratory rat), the rhythm peaks during the night; in diurnal animals (such as the tree shrew), it peaks during the day. In some animals, the rhythm is relatively symmetrical (for instance, the fat-tailed gerbil), while in others it has a complex shape (as in the tree shrew). Although the amplitude (excursion) of the rhythm sometimes depends on the environmental temperature under which the animal is kept,[72] different species display different amplitudes when maintained under the same constant environmental temperature (for instance, an amplitude of less than 2°C in the rat and more than 4°C in the tree shrew, as seen in Figure 3). The *human* body temperature rhythm is smooth and symmetrical, with a trough usually between 4 AM and 6 AM, a peak between 4 pm and 6 pm, and an amplitude of approximately 1 °C (2°F) around a daily mean of 37°C (98.6°F).[73-76]

An interesting feature of the body temperature rhythm (and of other rhythms as well) is that the daily rise in temperature seems to precede the transition between the two phases of the light–dark cycle. Thus, in Figure 3, the temperature of the gerbil starts to rise before the lights go off each day; similarly, the temperature of the tree shrew starts to rise before the lights go on each day. This means that somehow the animals anticipate the transition in the light-dark cycle. In order to do so consistently day after day, they have to possess some form of timing device. This timing device could be a simple "timer" (an hourglass) that is reset each day by the light-dark cycle, or it could be an actual "clock" that keeps time on its own. How can we tell which of the alternatives is the correct one? Some thinking will help: if a timer is reset daily by the light-dark cycle, then the anticipation will disappear if the light–dark cycle is eliminated. If daily rhythmicity persists in the absence of the light–dark cycle, then the clock alternative would seem to be correct. Can we test that? We certainly can. It has been done countless times. Daily rhythms of various biological variables have been found to persist in constant darkness or constant light in unicellular organisms,[41]

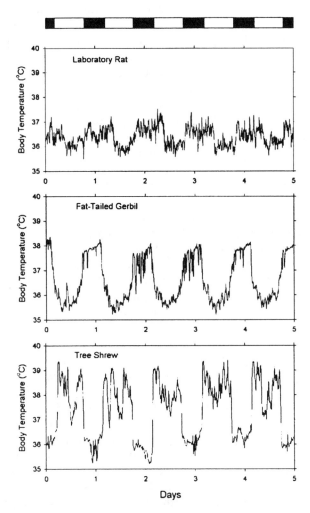

Figure 3. Five-day sections of the records of body temperature of representative individuals from three different species. All animals were maintained under an environmental temperature of 24°C (75°F). The light and dark bars at the top of the figure indicate the light and dark phases of the light–dark cycle.

mollusks,[77] crustaceans,[78] insects,[42,79–81] reptiles,[10,82,83] birds,[11,14,84–87] and mammals,[17,21,27,37,52,66,88–99] including humans.[100–107]

Of course, persistence of rhythmicity in the absence of a light-dark cycle does not prove that animals have an internal clock. They might very well have a timer that is reset not by light but by any of millions of geophysical variables that oscillate daily as the earth revolves around its axis (such as ambient temperature, humidity, magnetic fields, etc.). But then, how can we make sure that absolute constant conditions have been established in our study? We cannot. Does this mean that we cannot truly tell whether there is a real biological clock? Well, actually, luck is on our side. It happens that the rhythms recorded from animals maintained in constant darkness in the absence of other obvious time clues (such as cycles of ambient temperature or food availability) do not repeat themselves exactly every 24 hours. The *period* of these rhythms is almost always slightly different from 24.0 hours (for instance, 23.5 hours in the laboratory mouse and 24.8 hours in humans). Therefore, the rhythms must be *free-running*, as uncontrolled environmental variables cannot be resetting the rhythms each day — if they were, the period of the rhythms would be exactly 24.0 hours, as determined by the period of the earth's revolution. For true skeptics, however, I am glad to report that rhythms have been recorded in humans kept in underground bunkers and caves,[101,103] as well as in space.[108] Also, free-running rhythms were recorded in people living in Arctic and Antarctic field camps, where earth's influence is just as strong but there is continuous sunlight throughout the summer and continuous darkness throughout the winter.[106–107]

Of course, not every daily rhythm is generated by an internal clock. For instance, many office workers experience a daily rhythm of stress — they are very tense and apprehensive during the day, when their bosses keep a close eye on their performance, but are quite relaxed in the evening, when they return to the safety of their homes. This rhythm is clearly determined by the work schedule and not by an internal clock. On the other hand, the rhythm of body temperature *is* generated by an internal clock, as it free-runs in the absence of social schedules, light-dark oscillations, day-night differences in ambient temperature, and so on.[73, 103, 104] Those daily rhythms that have been found to free-run in a constant environment are called *circadian* rhythms (from the Latin, *circa* = approximately, *dies* = a day). Accordingly, the clock that sets the pace of circadian rhythms is called the *circadian pacemaker*. Rhythms that are present only when there is an environmental cycle to drive them (or rhythms whose endogenous nature has not yet been demonstrated) are called simply daily rhythms — or sometimes nychthemeral rhythms (from the Greek, *nychthémeron* = the duration of a day).

Since a large number of biological variables exhibit circadian rhythmicity, it is not unreasonable to wonder whether there are many circadian clocks. Is there one circadian pacemaker or several pacemakers? Since the existence of multiple clocks would create a serious problem of synchronization, it is reasonable to expect the existence of a single clock. But, of course, what is reasonable is not always true. Let us examine three main pieces of evidence.

In the presence of external time cues, different circadian rhythms in the same individual may peak at different times. For example, in humans, the rhythm of blood concentration of the hormone melatonin peaks during the night,[106, 109] whereas the body temperature rhythm peaks during the day.[73–76] There could be a separate clock for each of these two rhythms, but separate clocks are not necessary at all. Since both rhythms exhibit a 24-hour period, they are perfectly synchronized, even though they peak at different times.

A single clock can easily control the rise of the melatonin rhythm at the same time as the fall of the body temperature rhythm.

When animals (including humans) are subjected to an abrupt shift of the light–dark cycle, all rhythms must shift to maintain synchrony. However, different rhythms shift at different rates, thus creating a transitory state of internal desynchronization. For instance, in rats subjected to a 12-hour shift of the light–dark cycle, the body temperature rhythm readjusted in 9 days, while the rhythms of drinking and feeding readjusted in 7 days.[110] In humans subjected to a 6-hour shift, the body temperature rhythm readjusted in 6 days, whereas the activity rhythm readjusted in 2 days.[111] This means that, for a few days, different rhythms may exhibit different periods. It is possible, therefore, that separate clocks are responsible for the different periods. However, the different periods are clearly transient phenomena and are likely due to differential delays in the expression of the state of a single clock to the various organs (this will be discussed in greater detail in the next chapter in connection with the topic of resetting of the circadian pacemaker). After all, if there were multiple clocks, they would give stronger signs of their existence when allowed to free-run in constant conditions than when disturbed by a temporary event such as the shift in the light–dark cycle.

And this brings us to the third piece of evidence. In the majority of studies on free-running animals, all recorded rhythms were found to free-run together with the same period. There is always some amount of "biological noise" that could lead a hasty observer to see a transient episode of "spontaneous" internal desynchronization,[14,112] but, in the long run, no evidence of spontaneous internal desynchronization has ever been found in studies of non-human vertebrates maintained in constant darkness or constant light for extended periods of time.[52,88,89, 97,113,114] The same is true regarding many studies in humans,[104,106,115] but not all. Various investigators, in different laboratories, have found evidence of spontaneous internal desynchronization in human subjects maintained in constant conditions. Some have detected what they believed to be evidence for three distinct clocks,[116,117] but their results were never replicated and have been rightfully ignored. More commonly, spontaneous internal desynchronization has been found between the sleep-wake rhythm, on one side, and all other rhythms, on the other side.[100,103,118–122] Typically, the period of the sleep-wake rhythm becomes very long (up to 30 or even 50 hours) after a week or so, whereas the period of the body temperature rhythm (and other rhythms) remains between 24 and 25 hours. This happens in 15% to 60% of the subjects, depending on the study.

Before discussing the implications of the phenomenon of spontaneous internal desynchronization, we must make sure that we understand the phenomenon itself. Suppose that you go into a cave where you have no way to tell what time of the day it is. If you stay there for several weeks, the only clock you can rely on is your own internal clock. So, although you do not realize it, you will start going to bed later and later each day. Assuming a circadian period of 25 hours, you will go to bed at 10 pm today, 11 pm tomorrow, midnight the day after tomorrow, and so on. Of course, you do not realize that this is happening because you have no way to tell what time of the day it is — as a matter of fact, some male subjects have the impression that their beard is growing faster in the cave, as the interval between daily shaves increases without their awareness. In any event, if you do not develop internal desynchronization, you will be going to bed at noon after two weeks in the cave. This will also be the time of your day when you feel sleepy. However, if you do develop internal desynchronization, your life will be a mess. You will be going to sleep in 30-hour intervals (or more), while your sleepiness, your hunger, your need to go to the bathroom, and so on will be cycling in 25-hour intervals. Because of the oddity of such an arrangement, several researchers have suggested that internal desynchronization is unlikely to result from the dissociation of two biological clocks and may be due simply to a lot of noise in the expression of the sleep-wake rhythm.[104,123,124] Indeed, in internally desynchronized subjects — just as in normal subjects — sleepiness is strongly associated with the trough of the body temperature rhythm,[104,118,121] which means that people feel like sleeping (and often do sleep) at the particular time of the circadian day when the body temperature rhythm is at its low point. Now, if the sleep–wake rhythm and the body temperature rhythm are connected in such way, then they are obviously not independent of each other. In other words, they do not free-run separately. And, if they do not free-run separately, they need not be generated by separate circadian clocks.

Naturally, the absence of the requirement for separate pacemakers does not rule out the possibility of the existence of separate pacemakers. Mathematical models have been developed to show that the data obtained in internal desynchronization studies can be explained by the action of two separate, but coupled oscillators.[125,126] However, models have also been developed to show that the data can be explained by the action of a single pacemaker.[127,128] In the absence of additional evidence, parsimony leads us to conclude that there is only one circadian pacemaker.

Experimental evidence in support of the two-pacemaker hypothesis seemed to surface when some animal studies indicated that destruction of the brain area believed to control circadian rhythmicity of activity did not eliminate circadian rhythmicity of body temperature.[129–131] However, the apparent evidence was later shown to result from an artifact in the analysis of rhythmicity.[132] But let's not put the cart before the horse. Before we can address the issue of the physical location of the circadian pacemaker (Chapter 8), we must first complete our analysis of the functional properties of the pacemaker as revealed by the study of circadian rhythms. This will be our task for the next three chapters.

bibliographic notes

1. Ruffieux, C., Marazzi, A., and Paccaud, F. (1992). The circadian rhythm of the perinatal mortality rate in Switzerland. *American Journal of Epidemiology* 135: 936-951.
2. Gallerani, M., Avato, F. M., dal Monte, D., Caracciolo, S., Fersini, C., and Manfredini, R. (1996). The time for suicide. *Psychological Medicine* 26: 867-870.
3. Zautcke, J. L., Morris, R. W., Koeningsberg, M., Carmody, T., Stein-Spencer, L., and Erickson, T. B. (1998). Assaults from penetrating trauma in the State of Illinois. *American Journal of Emergency Medicine* 16: 553-556.
4. Fromm, R. E., Levine, R. L., and Pepe, P. E. (1992). Circadian variation in the time of request for helicopter transport of cardiac patients. *Annals of Emergency Medicine* 21: 1196-1199.
5. Rosenberg, J., Pedersen, M. H., Ramsing, T., and Krhlet, H. (1992). Circadian variation in unexpected postoperative death. *British Journal of Surgery* 79: 1300-1302.
6. Tomioka, K., Uwozumi, K., and Matsumoto, N. (1997). Light cycles given during development affect freerunning period of circadian locomotor rhythm of *period* mutants in *Drosophila melanogaster*. *Journal of Insect Physiology* 43: 297-305.
7. Roberts, M. H. and Xie, X. (1996). Phase relationship between ocular and behavioral circadian rhythms in *Bulla gouldiana* exposed to different photoperiods. *Physiology and Behavior* 59: 703-708.
8. Crawshaw, L. I. (1974). Temperature selection and activity in the crayfish, *Orconectes immunis*. *Journal of Comparative Physiology* 95: 315-322.
9. Reynolds, W. W., Casterlin, M. E., Matthey, J. K., Millington, S. T., and Ostrowski, A. C. (1978). Diel patterns of preferred temperature and locomotor activity in the goldfish *Carassius auratus*. *Comparative Biochemistry and Physiology* 59A: 225-227.
10. Refinetti, R. and Susalka, S. J. (1997). Circadian rhythm of temperature selection in a nocturnal lizard. *Physiology and Behavior* 62: 331-336.
11. Aschoff, J. and von Saint Paul, U. (1973). Brain temperature as related to gross motor activity in the unanesthetized chicken. *Physiology and Behavior* 10: 529-533.
12. Oshima, I. and Ebihara, S. (1988). The measurement and analysis of circadian locomotor activity and body temperature rhythms by a computer-based system. *Physiology and Behavior* 43: 115-119.
13. Pohl, H. (1976). Proportional effect of light on entrained circadian rhythms of birds and mammals. *Journal of Comparative Physiology* 112: 103-108.
14. Ebihara, S. and Gwinner, E. (1992). Different circadian pacemakers control feeding and locomotor activity rhythms in European starlings. *Journal of Comparative Physiology A* 171: 63-67.
15. Kramer, K., Voss, H. P., Grimbergen, J., and Bast, A. (1998). Circadian rhythms of heart rate, body temperature, and locomotor activity in freely moving mice measured with radio telemetry. *Lab Animal* 27(8): 23-26.
16. Refinetti, R. (1997). Phase relationship of the body temperature and locomotor activity rhythms in free-running and entrained rats. *Biological Rhythm Research* 28: 19-24.

17. Lovegrove, B. G., Heldmeier, G., and Ruf, T. (1993). Circadian activity rhythms in colonies of "blind" molerats, *Cryptomys damarensis* (Bathyergidae). *South African Journal of Zoology* 28: 46-55.
18. Brown, C. M. and Refinetti, R. (1996). Daily rhythms of heat production, body temperature, and locomotor activity in golden hamsters. *Journal of Thermal Biology* 21: 227-230.
19. Jilge, B. (1985). The rhythm of food and water ingestion, faeces excretion and locomotor activity in the guinea pig. *Zeitschrift für Versuchstierkunde* 27: 215-225.
20. Folk, G. E. (1957). Twenty-four hour rhythms of mammals in a cold environment. *American Naturalist* 91: 153-166.
21. Kennedy, G. A., Hudson, R., and Armstrong, S. M. (1994). Circadian wheel running activity rhythms in two strains of domestic rabbits. *Physiology and Behavior* 55: 385-389.
22. Lancia, R. A., Dodge, W. E., and Larson, J. S. (1982). Winter activity patterns of two radio-marked beaver colonies. *Journal of Mammalogy* 63: 598-606.
23. Menezes, A. A. L., Moreira, L. F. S., Azevedo, C. V. M., Costa, S. F., and Castro, C. S. S. (1993). Behavioral rhythms in the captive common marmoset (*Callithrix jacchus*) under natural environmental conditions. *Brazilian Journal of Medical and Biological Research* 26: 741-745.
24. Tapp, W. N. and Natelson, B. H. (1989). Circadian rhythms and patterns of performance before and after simulated jet lag. *American Journal of Physiology* 257: R796-R803.
25. Sulzman, F. M., Fuller, C. A., and Moore-Ede, M. C. (1977). Environmental synchronizers of squirrel monkey circadian rhythms. *Journal of Applied Physiology* 43: 795-800.
26. Plaza, A. V., Granda, T. G., and Cachero, M. T. G. (1993). Circadian rhythms of food and water intake and urine excretion in diabetic rats. *Physiology and Behavior* 54: 665-670.
27. Zucker, I. (1971). Light-dark rhythms in rat eating and drinking behavior. *Physiology and Behavior* 6: 115-126.
28. Ikeda, M. and Inoué, S. (1998). Simultaneous recording of circadian rhythms of brain and intraperitoneal temperatures and locomotor activity and drinking activities in the rat. *Biological Rhythm Research* 29: 142-150.
29. Rismiller, P. D. and Heldmeier, G. (1991). Seasonal changes in daily metabolic patterns of *Lacerta viridis*. *Journal of Comparative Physiology B* 161: 482-488.
30. Chakraborty, S. C., Ross, L. G., and Ross, B. (1992). The effect of photoperiod on the resting metabolism of carp (*Cyprinus carpio*). *Comparative Biochemistry and Physiology* 101A: 77-82.
31. Henken, A. M., Brandsma, H. A., van der Hel, W., and Verstegen, M. W. A. (1993). Circadian rhythm in heat production of limit-fed growing pigs of several breeds kept at and below thermal neutrality. *Journal of Animal Science* 71: 1434-1440.
32. Winget, C. M., Card, D. H., and Hetherington, N. W. (1968). Circadian oscillations of deep-body temperature and heart rate in a primate (*Cebus albafrons*). *Aerospace Medicine* 39: 350-353.
33. Smith, T. L., Coleman, T. G., Stanek, K. A., and Murphy, W. R. (1987). Hemodynamic monitoring for 24 h in unanesthetized rats. *American Journal of Physiology* 253: H1335-H1341.

34. Sato, K., Chatani, F., and Sato, S. (1995). Circadian and short-term variabilities in blood pressure and heart rate measured by telemetry in rabbits and rats. *Journal of the Autonomic Nervous System* 54: 235-246.

35. Roelfsema, F., van der Heide, D., and Smeenk, D. (1980). Circadian rhythms of urinary electrolyte excretion in freely moving rats. *Life Sciences* 27: 2303-2309.

36. Wasan, K. M., Brunner, L. J., Berens, K. L., Meltzer, A. A., and Luke, D. R. (1989). Circadian assessment of lipids in the hyperphagic obese rat compared with lean litter-mates. *Chronobiology International* 6: 223-228.

37. Maywood, E. S., Hastings, M. H., Max, M., Ampleford, E., Menaker, M., and Loudon, A. S. I. (1993). Circadian and daily rhythms of melatonin in the blood and pineal gland of free-running and entrained Syrian hamsters. *Journal of Endocrinology* 136: 65-73.

38. Wilson, M. E. and Gordon, T. P. (1989). Short-day melatonin pattern advances puberty in seasonally-breeding rhesus monkeys. *Journal of Reproduction and Fertility* 86: 435-444.

39. Halberg, F., Zander, H. A., Houglum, M. W., and Mühlemann, H. R. (1954). Daily variations in tissue mitoses, blood eosinophils and rectal temperature of rats. *American Journal of Physiology* 177: 361-366.

40. Chaudhry, A. P., Halberg, F., Keenan, C. E., Harner, R. N., and Bittner, J. J. (1958). Daily rhythms in rectal temperature and in epithelial mitoses of hamster pinna and pouch. *Journal of Applied Physiology* 12: 221-224.

41. Roennenberg, T. and Morse, D. (1993). Two circadian oscillators in one cell. *Nature* 362: 362-364.

42. Perkins, M. (1931). Light of glow-worms. *Nature* 128: 905.

43. Regal, P. J. (1967). Voluntary hypothermia in reptiles. *Science* 155: 1551-1553.

44. Kavaliers, M. and Ralph, C. L. (1980). Pineal involvement in the control of behavioral thermoregulation of the white sucker, *Catostomus commersoni*. *Journal of Experimental Zoology* 212: 301-303.

45. Ostheim, J. (1992). Coping with food-limited conditions: Feeding behavior, temperature preference, and nocturnal hypothermia in pigeons. *Physiology and Behavior* 51: 353-361.

46. Chossat, C. (1843). Recherches expérimentales sur l'inanition. II. Des effets de l'inanition sur la chaleur animale. *Annales des Sciences Naturelles 2e Série* 20: 293-326.

47. Davy, J. (1845). On the temperature of man. *Philosophical Transactions of the Royal Society of London* 135: 319-333.

48. Ogle, W. (1866). On the diurnal variations in the temperature of the human body in health. *St. George's Hospital Reports* 1: 221-245.

49. Hawking, F., Lobban, M. C., Gammage, K., and Worms, M. J. (1971). Circadian rhythms (activity, temperature, urine and microfilariae) in dog, cat, hen, duck, Thamnomys and Gerbillus. *Journal of Interdisciplinary Cycle Research* 2: 455-473.

50. Rashotte, M. E., Basco, P. S., and Henderson, R. P. (1995). Daily cycles in body temperature, metabolic rate, and substrate utilization in pigeons: Influence of amount and timing of food consumption. *Physiology and Behavior* 57: 731-746.

51. Hildwein, G. and Kayser, C. (1970). Relation entre la température colonique et la consommation d'oxygène d'un Insectivore, le Tenrec, au cours du nycthémère. *Comptes Rendus des Séances de la Société de Biologie de Strasbourg* 164: 429-432.

52. Refinetti, R. and Menaker, M. (1992). Body temperature rhythm of the tree shrew, *Tupaia belangeri*. *Journal of Experimental Zoology* 263: 453-457.

53. Weinert, D. and Waterhouse, J. (1998). Diurnally changing effects of locomotor activity on body temperature in laboratory mice. *Physiology and Behavior* 63: 837-843.

54. Haim, A., Ellison, G. T. H., and Skinner, J. D. (1988). Thermoregulatory circadian rhythms in the pouched mouse (*Saccostomus campestris*). *Comparative Biochemistry and Physiology* 91A: 123-127.

55. Rubal, A., Choshniak, I., and Haim, A. (1992). Daily rhythms of metabolic rate and body temperature of two murids from extremely different habitats. *Chronobiology International* 9: 341-349.

56. Körtner, G. and Geiser, F. (1995). Body temperature rhythms and activity in reproductive *Antechinus* (Marsupialia). *Physiology and Behavior* 58: 31-36.

57. Rose, R. W., Swain, R., and Bryant, S. L. (1990). Body temperature rhythm and regulation in the Tasmanian bettong (*Bettongia gaimardi*). *Comparative Biochemistry and Physiology* 97A: 573-576.

58. Heusner, A. (1959). Variation nycthémérale de la température centrale chez le rat adapté à la neutralité thermique.*Comptes Rendus des Séances de la Société de Biologie de Strasbourg* 153: 1258-1261.

59. Lovegrove, B. G. and Heldmeier, G. (1994). The amplitude of circadian body temperature rhythms in three rodents (*Aethomys namaquensis, Thallomys paedulcus* and *Cryptomys damarensis*) along the arboreal-subterranean gradient. *Australian Journal of Zoology* 42: 65-78.

60. Refinetti, R. (1995). Rhythms of temperature selection and body temperature are out of phase in the golden hamster. *Behavioral Neuroscience* 109: 523-527.

61. Refinetti, R. (1996).Comparison of the body temperature rhythms of diurnal and nocturnal rodents. *Journal of Experimental Zoology* 275: 67-70.

62. Refinetti, R. (1998). Homeostatic and circadian control of body temperature in the fat-tailed gerbil. *Comparative Biochemistry and Physiology* 119A: 295-300.

63. Varosi, S. M., Brigmon, R. L., and Besch, E. L. (1990). A simplified telemetry system for monitoring body temperature in small animals. *Laboratory Animal Science* 40: 299-302.

64. Hayes, S. R. (1976). Daily activity and body temperature of the southern woodchuck, *Marmota monax monax*, in northwestern Arkansas. *Journal of Mammalogy* 57: 291-299.

65. Dyck, A. P. and MacArthur, R. A. (1992). Seasonal patterns of body temperature and activity in free-ranging beaver (*Castor canadensis*). *Canadian Journal of Zoology* 70: 1668-1672.

66. Kuwabara, N., Seki, K., and Aoki, K. (1986). Circadian, sleep and brain temperature rhythms in cats under sustained daily light-dark cycles and constant darkness. *Physiology and Behavior* 38: 283-289.

67. Rawson, R. O., Stolwijk, J. A. J., Graichen, H., and Abrams, R. (1965). Continuous radio telemetry of hypothalamic temperatures from unrestrained animals. *Journal of Applied Physiology* 20: 321-325.

68. Mohr, E. and Krzywanek, H. (1990). Variations of core-temperature rhythms in unrestrained sheep. *Physiology and Behavior* 48: 467-473.

69. Bligh, J. and Harthoorn, A. M. (1965). Continuous radiotelemetric records of the deep body temperature of some unrestrained African mammals under near-natural conditions. *Journal of Physiology* 176: 145-162.

70. Simpson, S. and Galbraith, J. J. (1906). Observations on the normal temperature of the monkey and its diurnal variation, and on the effect of changes in the daily routine on this variation. *Transactions of the Royal Society of Edinburgh* 45: 65-104.

71. Fuller, C. A. (1984). Circadian brain and body temperature rhythms in the squirrel monkey. *American Journal of Physiology* 246: R242-R246.

72. Refinetti, R. (1997). The effects of ambient temperature on the body temperature rhythm of rats, hamsters, gerbils, and tree shrews. *Journal of Thermal Biology* 22: 281-284.

73. Aschoff, J., Gerecke, U., and Wever, R. (1967). Phasenbeziehungen zwischen den circadianen Perioden der Aktivität und der Kerntemperatur beim Menschen. *Pflügers Archiv* 295: 173-183.

74. Scales, W. E., Vander, A. J., Brown, M. B., and Kluger, M. J. (1988). Human circadian rhythms in temperature, trace metals, and blood variables. *Journal of Applied Physiology* 65: 1840-1846.

75. Elliott, A. L., Mills, J. N., Minors, D. S., and Waterhouse, J. M. (1972). The effect of real and simulated time-zone shifts upon the circadian rhythms of body temperature, plasma 11-hydroxycorticosteroids, and renal excretion in human subjects. *Journal of Physiology* 221: 227-257.

76. Kräuchi, K. and Wirz-Justice, A. (1994). Circadian rhythm of heat production, heart rate, and skin and core temperature under unmasking conditions in men. *American Journal of Physiology* 267: R819-R829.

77. Page, T. L., Wassmer, G. T., Fletcher, J., and Block, G. D. (1997). Aftereffects of entrainment on the period of the pacemaker in the eye of the mollusk *Bulla gouldiana*. *Journal of Biological Rhythms* 12: 218-225.

78. Cloudsley-Thompson, J. L. (1952). Studies in diurnal rhythms. II. Changes in the physiological responses of the woodlouse *Oniscus asellus* to environmental stimuli. *Journal of Experimental Biology* 29: 295-303.

79. Moritz, R. F. A. and Sakofski, F. (1991). The role of the queen in circadian rhythms of honeybees (*Apis mellifera* L.). *Behavioral Ecology and Sociobiology* 29: 361-365.

80. White, L., Ringo, J., and Dowse, H. (1992). A circadian clock of *Drosophila*: Effects of deuterium oxide and mutations at the *period* locus. *Chronobiology International* 9: 250-259.

81. Takahashi-del-Bianco, M., Benedito-Silva, A. A., Hebling, M. J. A., Marques, N., and Marques, M. D. (1992). Circadian oscillatory patterns of oxygen uptake in individual workers of the ant *Camponotus rufipes*. *Physiological Entomology* 17: 377-383.

82. Hoffmann, K. (1968). Synchronisation der circadianen Aktivitätsperiodik von Eidechsen durch Temperaturcyclen verschiedener Amplitude. *Zeitschrift für Vergleichende Physiologie* 58: 225-228.

83. Cowgell, J. and Underwood, H. (1979). Behavioral thermoregulation in lizards: A circadian rhythm. *Journal of Experimental Zoology* 210: 189-194.

84. Graf, R. (1980). Diurnal changes of thermoregulatory functions in pigeons. I. Effector mechanisms. *Pflügers Archiv* 386: 173-179.

85. Pohl, H. (1993). Does aging affect the period of the circadian pacemaker in vertebrates? *Naturwissenschaften* 80: 478-481.

86. Eskin, A. (1971). Some properties of the system controlling the circadian activity rhythm of sparrows. In: Menaker, M. (Ed.). *Biochronometry*. Washington, DC: National Academy of Sciences, pp. 66-80.

87. Aschoff, J. and von Saint Paul, U. (1976). Brain temperature in the unanesthe-tized chicken: Its circadian rhythm of responsiveness to light. *Brain Research* 101: 1-9.

88. Honma, K. and Hiroshige, T. (1978). Simultaneous determination of circadian rhythms of locomotor activity and body temperature in the rat. *Japanese Journal of Physiology* 28: 159-169.

89. Eastman, C. and Rechtschaffen, A. (1983). Circadian temperature and wake rhythms of rats exposed to prolonged continuous illumination. *Physiology and Behavior* 31: 417-427.

90. McElhinny, T. L., Smale, L., and Holekamp, K. E. (1997). Patterns of body temperature, activity, and reproductive behavior in a tropical murid rodent, *Arvicanthis niloticus. Physiology and Behavior* 62: 91-96.

91. Lovegrove, B. G. and Papenfus, M. E. (1995). Circadian activity rhythms in the solitary cape mole rat (*Georychus capensis*: Bathyergidae) with some evi-dence of splitting. *Physiology and Behavior* 58: 679-685.

92. Berezkin, M. V., Kudinova, V. F., Batygov, A. N., Ponomareva, L. E., and Zhukova, G. N. (1989). Effect of lighting conditions on circadian rhythm of rectal temperature in mice. *Bulletin of Experimental Biology and Medicine* 106: 1337-1340.

93. Büttner, D. (1992). Social influences on the circadian rhythm of locomotor activity and food intake of guinea pigs. *Journal of Interdisciplinary Cycle Research* 23: 100-112.

94. Menaker, M. (1959). Endogenous rhythms of body temperature in hibernating bats. *Nature* 184: 1251-1252.

95. Recabarren, S. E., Vergara, M., Llanos, A. J., and Serón-Ferré, M. (1987). Cir-cadian variation of rectal temperature in newborn sheep. *Journal of Develop-mental Physiology* 9: 399-408.

96. Lowe, C. H., Hinds, D. S., Lardner, P. J., and Justice, K. E. (1967). Natural free-running period in vertebrate animal populations. *Science* 156: 531-534.

97. Refinetti, R. (1996). The body temperature rhythm of the thirteen-lined ground squirrel, *Spermophilus tridecemlineatus. Physiological Zoology* 69: 270-275.

98. Moore-Ede, M. C., Kass, D. A., and Herd, J. A. (1977). Transient circadian internal desynchronization after light-dark phase shift in monkeys. *American Journal of Physiology* 232: R31-R37.

99. Hoban, T. M., Levine, A. H., Shane, R. B., and Sulzman, F. M. (1985). Circadian rhythms of drinking and body temperature of the owl monkey (*Aotus trivir-gatus*). *Physiology and Behavior* 34: 513-518.

100. Aschoff, J., Gerecke, U., and Wever, R. (1967). Desynchronization of human circadian rhythms. *Japanese Journal of Physiology* 17: 450-457.

101. Colin, J., Timbal, J., Boutelier, C., Houdas, Y., and Siffre, M. (1968). Rhythm of rectal temperature during a 6-month free-running experiment. *Journal of Applied Physiology* 25: 170-176.

102. Kriebel, J. (1974). Changes in internal phase relationships during isolation. In: Scheving, L. E., Halberg, F., and Pauly, J. E. (Eds.). *Chronobiology.* Tokyo: Igaku Shoin, pp. 451-459.

103. Lund, R. (1974). Personality factors and desynchronization of circadian rhythms. *Psychosomatic Medicine* 36: 224-228.

104. Zulley, J. and Campbell, S. S. (1985). Napping behavior during "spontaneous internal desynchronization": Sleep remains in synchrony with body temper-ature. *Human Neurobiology* 4: 123-126.

105. Pollak, C. P. and Wagner, D. R. (1994). Core body temperature in narcoleptic and normal subjects living in temporal isolation. *Pharmacology Biochemistry and Behavior* 47: 65-71.
106. Kennaway, D. J. and van Dorp, C. F. (1991). Free-running rhythms of melatonin, cortisol, electrolytes, and sleep in humans in Antarctica. *American Journal of Physiology* 260: R1137-R1144.
107. Steel, G. D., Callaway, M., Suedfeld, P., and Palinkas, L. (1995). Human sleep-wake cycles in the high Arctic: Effects of unusual photoperiodicity in a natural setting. *Biological Rhythm Research* 26: 582-592.
108. Gundel, A., Polyakov, V. V., and Zulley, J. (1997). The alteration of human sleep and circadian rhythms during spaceflight. *Journal of Sleep Research* 6: 1-8.
109. Lewy, A. J. and Sack, R. L. (1989). The dim light melatonin onset as a marker for circadian phase position. *Chronobiology International* 6: 93-102.
110. Zerath, E., Holy, X, Lagarde, D., Fernandes, T., Rousselet, D., and Lalouette, A. (1994). Dissociation in body temperature, drinking and feeding rhythms following a light-dark cycle inversion in the rat. *Medical Science Research* 22: 53-55.
111. Wever, R. A. (1980). Phase shifts of human circadian rhythms due to shifts of artificial zeitgebers. *Chronobiologia* 7: 303-327.
112. Sulzman, F. M., Fuller, C. A., and Moore-Ede, M. C. (1977). Spontaneous internal desynchronization of circadian rhythms in the squirrel monkey. *Comparative Biochemistry and Physiology* 58A: 63-67.
113. Honma, K. and Hiroshige, T. (1978). Internal synchronization among several circadian rhythms in rats under constant light. *American Journal of Physiology* 235: R243-R249.
114. Fuller, C. A. and Edgar, D. M. (1986). Effects of light intensity on the circadian temperature and feeding rhythms in the squirrel monkey. *Physiology and Behavior* 36: 687-691.
115. Campbell, S. S., Dawson, D., and Zulley, J. (1993). When the human circadian system is caught napping: Evidence for endogenous rhythms close to 24 hours. *Sleep* 16: 638-640.
116. Folkard, S., Wever, R. A., and Wildgruber, C. M. (1983). Multi-oscillatory control of circadian rhythms in human performance. *Nature* 305: 223-226.
117. Folkard, S., Hume, K. I., Minors, D. S., Waterhouse, J. M., and Watson, F. L. (1985). Independence of the circadian rhythm in alertness from the sleep/wake cycle. *Nature* 313: 678-679.
118. Zulley, J., Wever, R., and Aschoff, J. (1981). The dependence of onset and duration of sleep on the circadian rhythm of rectal temperature. *Pflügers Archiv* 391: 314-318.
119. Aschoff, J. (1994). Naps as integral parts of the wake time within the human sleep-wake cycle. *Journal of Biological Rhythms* 9: 145-155.
120. Wever, R. A. (1989). Light effects on human circadian rhythms: A review of recent Andechs experiments. *Journal of Biological Rhythms* 4: 161-185.
121. Czeisler, C. A., Weitzman, E. D., Moore-Ede, M. C., Zimmerman, J. C., and Knauer, R. S. (1980). Human sleep: Its duration and organization depend on its circadian phase. *Science* 210: 1264-1267.
122. Green, J., Pollak, C. P., and Smith, G. P. (1987). Meal size and internal interval in human subjects in time isolation. *Physiology and Behavior* 41: 141-147.

123. Mills, J. N. (1973). Transmission processes between clock and manifestations. In: Mills, J. N. (Ed.). *Biological Aspects of Circadian Rhythms.* New York: Plenum, pp. 27-84.

124. Minors, D. S., Waterhouse, J. M., Folkard, S., and Totterdell, P. (1993). Circadian rhythms of body temperature and urinary excretion in subjects on 30-h "days." *Journal of Interdisciplinary Cycle Research* 24: 277-283.

125. Aschoff, J. and Wever, R. (1976). Human circadian rhythms: A multioscillatory system. *Federation Proceedings* 35: 2326-2332.

126. Kronauer, R. E., Czeisler, C. A., Pilato, S. F., Moore-Ede, M. C., and Weitzman, E. D. (1982). Mathematical model of the human circadian system with two interacting oscillators. *American Journal of Physiology* 242: R3-R17.

127. Eastman, C. (1984). Are separate temperature and activity oscillators necessary to explain the phenomena of human circadian rhythms? In: Moore-Ede, M. C. and Czeisler, C. A. (Eds.). *Mathematical Models of the Circadian Sleep-Wake Cycle.* New York: Raven, pp. 81-103.

128. Daan, S., Beersma, D.G.M., and Borbély, A.A. (1984). Timing of human sleep: Recovery process gated by a circadian pacemaker. *American Journal of Physiology* 246: R161-R178.

129. Dunn, J. D., Castro, A. J., and McNulty, J. A. (1977). Effect of suprachiasmatic ablation on the daily temperature rhythm. *Neuroscience Letters* 6: 345-348.

130. Fuller, C.A., Lydic, R., Sulzman, F.M., Albers, H.E., Tepper, B., and Moore-Ede, M.C. (1981). Circadian rhythm of body temperature persists after suprachiasmatic lesions in the squirrel monkey. *American Journal of Physiology* 241: R385-R391.

131. Satinoff, E. and Prosser, R.A. (1988). Suprachiasmatic nuclear lesions eliminate circadian rhythms of drinking and activity but not of body temperature in male rats. *Journal of Biological Rhythms* 3: 1-22.

132. Refinetti, R., Kaufman, C. M., and Menaker, M. (1994). Complete suprachiasmatic lesions eliminate circadian rhythmicity of body temperature and locomotor activity in golden hamsters. *Journal of Comparative Physiology A* 175: 223-232.

chapter five

the operation of the clock

In this chapter, the operation of the circadian pacemaker
is described in detail. This includes a discussion of
genetic influences and mechanisms by which the clock
interacts with environmental stimuli.

Although daily and circadian rhythms in plants and animals have been
occasionally studied for centuries, multiple systematic studies did not start
until the 1950s, when three independent researchers laid down the foundations of modern circadian biology. The three "fathers" of circadian biology
were: Jürgen Aschoff (at the Max Planck Institute, in Andechs, Germany),
Colin Pittendrigh (first at Princeton University, in New Jersey, and later at
Stanford University, in California), and Franz Halberg (at the University of
Minnesota). Ironically, Dr. Halberg was the creator of the term *circadian* but
conducted little research on free-running rhythms. Most of his work concentrated on the relationships between daily rhythms in humans and animals
living normal lives. He was an ardent proponent of chronopharmacology—the study and implementation of optimal schedules to administer
medicine in the light of what is known about circadian variations in drug
absorption, drug metabolism, and tissue sensitivity. Dr. Pittendrigh, on the
other hand, was interested in the basic mechanisms of biological timing.
With a background in physics, he strove to understand how the operation
of a physical oscillator could explain circadian rhythmicity in animals. Most
of our current understanding of the operation of the circadian clock is
derived from his work. Last, but not least, Dr. Aschoff was a true naturalist.
He was interested in all manifestations of circadian rhythmicity, in the laboratory as well as in the field. A prolific researcher, he investigated a wide
variety of phenomena in a multitude of species, including humans. His
discovery and interpretation of the phenomenon of spontaneous internal
desynchronization was a driving force in circadian biology for decades. His
thorough and exhaustive reviews of the literature in circadian biology served
as invaluable guides to numerous researchers. I was most honored when, in
a personal letter to me, Dr. Aschoff referred to a review article of mine[1] as

a "masterpiece." I have also had the honor and pleasure of meeting the late Drs. Aschoff and Pittendrigh at scientific meetings.

the endogenous clock

In the preceding chapter, we examined the massive evidence that an internal clock (the circadian pacemaker) is responsible for the generation of circadian rhythms. Under constant environmental conditions, the rhythms free-run with periods slightly different from 24.0 hours. Naturally, the endogenous nature of the clock necessitates some form of genetic inheritance. Since different species have clearly distinct circadian periods, even though they live in similar environments, genetic influences must be stronger than environmental influences. However, additional information is necessary to ascertain how strong the genetic influence is. Valuable information has been provided by various studies dealing with mutations of the genes that control circadian rhythmicity. In the early 1970s, specific single-gene mutations were identified in the fruit fly, *Drosophila melanogaster,*[2] and in the bread mold, *Neurospora crassa.*[3] In the fruit fly, the gene — called *per*, for period — is located in the X chromosome and exhibits multiple mutant alleles. The first three alleles that were identified shortened (*per^S*, $\tau \approx 19$ hours), lengthened (*per^L*, $\tau \approx 29$ hours), or eliminated (*per^0*) circadian rhythmicity of activity. Thus, a homozygous fly with the *per^S* alleles maintained in a constant environment lived a 19-hour day rather than a 24-hour day! "Flies from Mars!"— could have been the newspaper headline, except that the Martian day is actually very close to 24 hours ("Flies from Neptune!" would be a little more accurate, although still incorrect). Of course, the fact that a single gene mutation can have such a drastic effect on free-running rhythms indicates that the period of the circadian pacemaker has a very strong genetic component, at least in fruit flies and fungi. In 1989, a single-gene mutation was identified in the golden hamster,[4] and in 1994 a mutation was produced in the laboratory mouse by mutagenesis and screening.[5]

While the mutation in the mouse (the *Clock* gene, located in chromosome 5) is better understood at the molecular level than the mutation in the golden hamster (see chapter 10), the behavioral and physiological effects of the mutation in the hamster (the *tau* gene) have been studied more extensively. The discovery of the mutation was fortuitous. Martin Ralph, now a professor at the University of Toronto, was then a graduate student in Dr. Michael Menaker's laboratory at the University of Oregon. As part of his experiments, Ralph had numerous hamsters free-running in constant darkness. One of the hamsters displayed an unusual pattern of activity, seeming to phase advance 2 hours each day. Knowing about the studies in fruit flies, Ralph restrained himself from the impulse to discard the "defective" hamster and replace it with a "good" one. Rather, he started inbreeding successive generations of the unusual hamster. He soon determined that the abnormal behavior was the result of a single-gene mutation with a single mutant allele, *tau*. While normal hamsters (+/+) have a circadian period very close to 24 hours, heterozygous mutants (+/*tau*) have a period of 22 hours, and homozygotes (*tau*/*tau*) have a period of 20 hours.[4] "Hamsters from Mars!"

When I joined Dr. Menaker's laboratory as a postdoctoral fellow in 1990, he had moved to the University of Virginia and had brought the mutants with him. Since the *tau* mutation had only been shown to affect the period of the activity rhythm, I was curious to know whether it affected also the body temperature rhythm. In my studies, I found out that the mutation did affect the temperature rhythm, and it did so in a very elegant manner — by proportionally shortening the rhythm without affecting its shape or its mean level[6] — which reinforced the assumption that the effect of the gene was general to the circadian pacemaker rather than restricted to a particular effector mechanism. I also wanted to be sure that the mutation was specific to the circadian system and did not alter other processes of biological timing. Through further research, I obtained evidence that the *tau* mutation did not affect the frequency of the pacemaker of the heart[7] or the duration of the estrous cycle[8] (the equivalent of the menstrual cycle in rodents). Years later, as a professor working in my own laboratory at the College of William & Mary in Williamsburg, Virginia, I also investigated ultradian[9] and metabolic[10] alterations caused by the *tau* mutation. Dr. Andrew Loudon, at the University of Manchester in England, recently showed that the *tau* mutation proportionally shortens the rhythms of melatonin, luteinizing hormone, and cortisol secretion.[11]

So, the circadian period seems to have a very strong genetic component. But does the early environment — during development of the adult organism — have any effect at all? Pittendrigh and other investigators had noticed that the period of free-running rhythms in adults can be affected by the previous exposure of the animal to light-dark cycles with different periods.[12-14] For instance, animals previously maintained under a light-dark cycle with a 22-hour period have a shorter circadian period (when allowed to free-run in constant darkness) than animals previously maintained under a regular, 24-hour day. This effect may last for months but eventually disappears. Would these "after-effects" become permanent if the animals were raised from birth (or, even better, from conception) under different light-dark cycles? As far as vertebrates are concerned, this question has not been addressed experimentally often enough to allow a definitive answer, but two studies in mice[15] and hamsters[16] have revealed absolutely no effect of the early environment beyond that expected as an after-effect of the light-dark cycle (as observed in adults). Also, regardless of the exact period of the light-dark cycle, the presence of a light–dark cycle per se is not necessary to "prime" the operation of the circadian clock, as animals raised in constant darkness or constant light since birth exhibit normal free-running rhythms as adults.[17-19] It seems, therefore, that the old nature–nurture controversy has been decided in favor of nature as far as the period of the circadian pacemaker in vertebrates is concerned (the situation may be different in invertebrates[20]).

Still, we have not addressed the fundamental question: How does the circadian clock "generate" time? The answer to this question is both extremely simple and extremely complex. Since all biochemical reactions take place over time, time in itself is inherent to any biological process. All that is needed to produce a biological clock is a biochemical loop, that is, a series of reactions that repeats itself under a constant rate. Several simple models of how this could be accomplished have been proposed.[21] If, for instance, structure A produces substance B, and substance B feedbacks on structure A to inhibit its own production, a biological clock will exist and will have a period equal to the time needed to produce enough substance B to reach the inhibitory threshold plus the time needed for substance B to be metabolized down to the threshold concentration. Most likely, the real process involves numerous steps, but the concept of a biological clock in itself is rather simple and requires no magical elements. While the actual biochemical reactions responsible for the circadian pacemaker are still not known, great progress has been achieved recently, as will be discussed in Chapter 10.

entrainment

Fascinating as the endogenous nature of circadian rhythms may be, the truth is that most animals in natural environments are not free-running. The presence of daily cycles of light and darkness, ambient temperature, and food availability is the rule, not the exception. How does the circadian pacemaker adapt to these cycles? Before we attempt to answer this question, we need to agree on some terminology. Clearly, our interest is in understanding how the period of the circadian pacemaker (symbolized by the Greek lower case letter tau, τ) comes to match the period of an external clock (symbolized by the Roman upper case letter T). Since the external clock gives the time to the internal clock, it can be called a "time giver" (or, as originally named in German, a *Zeitgeber*). If the zeitgeber is effective, the internal clock will go along with it, as if it had boarded a train; thus, we will say that the pacemaker was *entrained* by the zeitgeber. And, in possession of this terminology, you can impress all your friends by casually stating that "the entrainment of the circadian pacemaker by a zeitgeber is attained by the modulation of τ toward T." In case you yourself need some time to make sense out of this sentence, we can reword it as: "the biological clock synchronizes to the outside world by adjusting to a 24-hour day."

So that we can visualize changes in τ, we need a convenient way to graph the temporal course of circadian rhythms. The traditional graph in circadian biology is the *actogram*, which was developed before the graphical advances of the personal computer age but has since incorporated the new technology. As shown in Figure 4, the procedure starts with raw data obtained by the continuous recording of locomotor activity. The use of raw data has two disadvantages: (1) each day is plotted to the right of the preceding day, which makes it difficult to compare the temporal distribution of activity on different days, and (2) the temporal resolution of the plot is rather low, as several days are plotted on the same line. These two disadvantages can be overcome by "cutting out" the data for each day, "stretching" it to the full width of a page, and "pasting" it below the preceding day. The resulting graph shows data for one day per line, with successive days appearing on successive lines. The vertical alignment of the data provides instant information about the period of the rhythm: inclination to the left indicates a period shorter than 24 hours (onsets are earlier each day) whereas inclination to the right indicates a period longer than 24 hours (onsets are later each day). The inclination to the right in Figure 4 indicates that the period of the rhythm is longer than 24 hours.

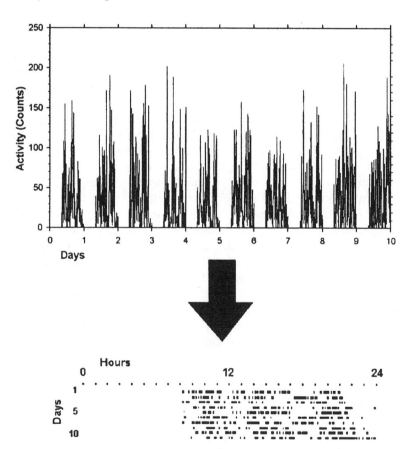

Figure 4. A 10-day segment of the records of locomotor activity of a rat maintained in constant darkness and the actogram derived from that segment. Notice that information about absolute amounts of activity is lost in the actogram but that the temporal distribution of activity can be much more easily observed.

Now, back to the notion of entrainment. If the pacemaker is to be entrained by the zeitgeber (or, as it is sometimes said, if the pacemaker is to entrain to the zeitgeber), then its free-running period (τ) must be changed to match that of the zeitgeber (T). From extensive experimentation with man-made clocks, we know that there are two ways to change the period of a clock. We can actually change the speed of the clock, thus making it run faster or slower, or we can simulate a change in speed by resetting the clock each day. If you have one of those old winding watches, you must have employed the latter procedure recently. If the watch is slow (that is, it takes longer than 24 hours for the small arm to go around the dial), you can fix it by advancing the arms a little bit each day. On the other hand, if the watch is fast, you can fix it by moving back (or "delaying") the arms a little bit each day. To use Pittendrigh's terminology,[22] the amount of time that needs to be adjusted each day (that is, the *shift* in the clock mechanism, symbolized by delta phi, $\Delta\phi$) is given by a simple equation: $\Delta\phi = \tau - T$ ("delta *fea* equals *tau* minus *tea*"). Thus, if the period of the clock is 23.8 hours (a fast clock), you need to delay it by 12 minutes ($\Delta\phi = -0.2$ hour) each day; if, on the other hand, the period is 24.1 hours (a slow clock), you need to advance the clock by 6 minutes ($\Delta\phi = +0.1$ hour) each day. Notice that, given the arrangement of the equation, delays are represented by negative values and advances by positive values. Of course, this is an arbitrary arrangement; however, it has become a standard convention just like that of representing a positive event by the number 1 and a null event by the number 0 in computer operations.

You might guess that I would not be talking about $\Delta\phi$ if entrainment were due to the slowing down or speeding up of the circadian pacemaker. You are right. According to Pittendrigh's so-called "non-parametric theory of entrainment," entrainment is due to the daily resetting of the pacemaker by the zeitgeber. In contrast, "parametric" theories of entrainment are based on the assumption that the magnitude of the zeitgeber has a proportional (parametric) effect on the speed of the clock. The non-parametric theory has been so successful in explaining experimental data that very few researchers today even think about parametric theories. A short tutorial on phase reset-ting is included in the software package that accompanies this book (program Entra in.exe).

Because most of the fundamental research on entrainment has been conducted using photic stimuli (light) as the zeitgeber, I will limit all discus-sions in this chapter to this class of stimuli. Non-photic zeitgebers will be discussed in Chapter 6.

Since animals maintained in constant darkness are not exposed to light, the period of their rhythms under this condition can be considered to be the expression of the natural (uncontaminated) period of the circadian pace-maker. In constant light, the photic stimulation does not provide time cues, but light may (and actually does) affect the operation of the pacemaker. In most nocturnal animals, τ lengthens as the intensity of (constant) light increases, whereas in most diurnal animals it shortens as light intensity increases (Pittendrigh called this Aschoff's rule,[23] after Aschoff's observation of the relationship in numerous publications). As discussed later on, this apparent parametric effect of light can be explained by non-parametric mechanisms (which Aschoff himself acknowledged[24]). In any event, it should be pointed out that the effect of light intensity on the circadian period measured in constant light tells us very little about entrainment, as animals do not entrain to constant light (because constant light is as devoid of time cues as constant darkness is).

How does light affect the circadian pacemaker? Pittendrigh's original work used the eclosion of fruit fly pupae as the variable of measurement, but he later replicated his studies using the locomotor activity rhythm in rodents. His fundamental finding was that, in animals maintained in constant darkness, brief light pulses cause consistent shifts in the phase of the activity rhythm (and, since the rhythm expresses the state of the clock, the phase shift in the rhythm corresponds to the resetting of the clock). The magnitude and direction of the shifts depend on the time (relative to the animal's clock) at which the pulse is presented. In most mammals, pulses presented during early "subjective night" (that is, the time in the animal's clock that would correspond to darkness if a light–dark cycle were present) cause phase delays, pulses presented during late subjective night cause advances, and pulses presented during most of subjective day usually have no effect.[12] Thus, there is a *phase-response curve* (PRC) that relates the phase shifts in the rhythm to the time of light presentation. Phase-response curves have been determined for many species[12,25–32] and have been found to be similar in general features but to differ considerably in amplitude and exact timing from one species to another. Light pulses have also been shown to phase-shift human rhythms,[33–39] and preliminary human PRCs have been published.[40,41]

Because of the importance of the phase-response curve for the process of entrainment, we must discuss it in some detail. To make things easier, let us assume an animal with $\tau = 24$ hours. In this case, the animal's clock is similar to our man-made clocks, in the sense that each hour of the animal's clock corresponds to 1/24 of the period of the rotation of the earth around its axis. Thus, "circadian time" (CT) is the same as actual time. If we define "circadian time 12" (CT 12) as the time at which the animal becomes active each day, then subjective night starts at CT 12, assuming a nocturnal species (such as the golden hamster). To avoid an inconsistency in labeling, the onset of activity is defined as CT 0 in diurnal animals, so that the interval between CT 12 and the end of the circadian day still refers (approximately) to subjective night. If $\tau \neq 24$ hours, everything remains the same, except that each circadian hour lasts $\tau/24$ of a physical hour.

Figure 5 shows some actograms depicting phase shifts of the activity rhythm of golden hamsters caused by light pulses presented at different circadian times. For instance, a pulse presented at CT 2 causes no detectable shift, a pulse at CT 14 causes a phase delay of about an hour, and a pulse presented at CT 18 causes a phase advance of over 2 hours. Figure 5 also

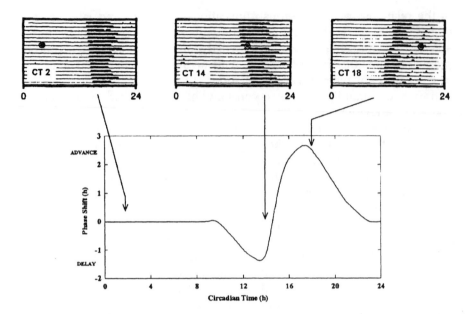

Figure 5. Representative actograms and the full hamster phase–response curve for brief light pulses presented a week after transfer of the animals from a light–dark cycle to constant darkness. The concentric circles in the actograms indicate the time of the light pulses.

indicates how the phase shifts are plotted as a function of circadian time to generate a phase-response curve.

As mentioned above, phase-response curves are species specific, and the hamster PRC should not be assumed to predict the human PRC. Studies in humans have often suffered from methodological limitations — in particular, several investigators have used short "constant routine" protocols to free the circadian pacemaker from environmental influences, which is a clever approach but does not provide the opportunity for rigorous data analysis afforded by long free-runs in constant darkness. Nevertheless, it seems clear that the human PRC differs from the hamster PRC. For instance, in humans, advances of 3 to 4 hours are attained by pulses presented during early subjective day;[40–41] in hamsters, the largest advance is only about 2.4 hours and is evoked during subjective night (Figure 5).

Once you understand that light pulses cause phase shifts, it is very easy to understand entrainment. Let us say that an animal with a free-running period of 23.5 hours must entrain to a 24 hour cycle. If $\Delta\phi = \tau - T$, then the animal must delay 30 minutes each day (23.5 − 24.0 = -0.5 hour). Using the PRC in Figure 5, we can see that a 30-minute delay is caused by a light pulse at CT 11. Thus, if we present a light pulse at CT 11 each day, we will cause a 30-minute delay each day, and the observed period of the activity rhythm will be 24 hours. That is, the animal will be entrained. As a matter of fact, we do not even need to know when CT 11 is. If we present pulses every 24 hours, the animal's rhythm will be initially advanced or delayed (or not affected) haphazardly as light "hits" the pacemaker at different circadian times each cycle, until the point is reached at which the daily light pulse causes the exact 30-minute delay necessary for entrainment — at this point, the animal will be "locked" into the entrained position.

Of course, if you are a detailed observer, you noticed that, to continue with our example, a 30-minute delay can be caused not only by a pulse at CT 11 but also at CT 15. However, the proximity between CT 15 and the beginning of the advance zone of the PRC makes entrainment at this position rather unstable. Also, as a detailed observer, you noticed that there must be a natural *range of entrainment* — that is, a limited range of periods of the zeitgeber to which the pacemaker can entrain. Since the maximum delay in the hamster PRC (Figure 5) is one hour, hamsters should be unable to entrain to a light-dark cycle with a period longer than 25.1 hours (that is, 1 hour more than the free-running period of 24.1 hours). Similarly, entrainment should not be possible to light-dark cycles shorter than 21.7 hours. Empirical studies of the range of entrainment have been carried out in various species,[13,14,32,42,43] but they have rarely examined in detail the relationship with the amplitude of the PRC.

Now, if you not only are a detailed observer but also have a great ability to infer logical consequences, you may have come up with the concept of entrainment by *frequency demultiplication.*[22] That is, the range of entrainment can be extended if multiples of τ are used as T. Indeed, if light pulses are presented every 48 hours (T = 48.0 hours), they will still be able to produce phase shifts large enough to compensate for the spontaneous drift of the pacemaker. To keep using the hamster example, a 0.2-hour phase-advance every other day will be just as effective as a 0.1-hour advance each day. And, as long as light does not start hitting the system at multiple points of the PRC, zeitgeber periods much shorter than 24 hours can also cause entrainment. For example, light pulses presented every 12 hours (T = 12.0 hours) can produce entrainment by hitting the system at the phase advance region and the dead zone each cycle.

So, light pulses can cause phase shifts, and phase shifts are theoretically sufficient to account for entrainment. But has this actually been tested? Yes. Laboratory studies using daily light pulses as entraining agents have been conducted and have confirmed that stable entrainment can be attained with the presentation of brief daily light pulses.[44-46] Even more interesting were studies using rodents allowed to display their natural behavioral pattern of spending most of the time in underground burrows. Since these animals only come out of the burrow at night, they are rarely exposed to the light-dark cycle. They actually get exposed only to brief pulses of light every few days. In these cases, entrainment is clearly the result of discrete phase shifts interspersed with several days of free-run[47,48] (entrainment by frequency demultiplication).

When a full light-dark cycle is present, entrainment is only slightly complicated by the presence of additional, "unnecessary" light. Most of the extra light is left in the insensitive zone of the PRC (to the left of the phase-delay region when τ is shorter than T, and to the right of the phase-advance region when τ is longer than T). However, more than a single point of the phase-delay or phase-advance regions will be stimulated. The net effect will presumably correspond to the integration of the area under the curve in the stimulated range. Although manual computation of the area under the curve can be extremely tedious, computer programs can calculate it easily. Several computer models have been developed.[29,49,50] The outcome of the simulations is consistent with the actual behavior of animals exposed to full light-dark cycles, except for one unexpected element. Bizarre as it may seem, simulations based exclusively on the known action of light on the PRC indicate that animals cannot free-run in constant light — which is clearly not true. This can be corrected by assuming that, just like in the visual system, the circadian photic system is subject to light adaptation (that is, photic sensitivity is temporarily reduced by exposure to light, which you have certainly experienced when going into a dark movie theater after being under bright sunlight outside). This correction actually results in entrainment simulations much more consistent with the entrainment of actual animals.[23,50]

Surprisingly, there have been very few studies of light adaptation in the circadian system, but they do support the assumptions of the models.[51, 52] For instance, a light pulse presented at CT 13 to a hamster maintained in constant darkness for the "standard" interval of one week causes a phase delay of about 1 hour (Figure 5). However, if the animal is maintained in constant darkness for a full month, the same light pulse causes a delay of 2 hours. Indirect evidence concerning light adaptation in the circadian system comes also from studies in animals maintained under light-dark cycles with long light phases (16 hours of light per day) or short light phases (8 hours of light per day). In animals previously maintained under a cycle with short light phase (short photoperiod), the sensitive zone of the PRC is 14 hours wide, whereas the sensitive zone is only 8 hours wide in animals previously maintained under a cycle with long light phase.[53, 54] Presumably, the reduced width of the sensitive zone of the PRC is a result of a reduction in photic sensitivity due to the greater pre-exposure to light in animals maintained under light–dark cycles with long light phases.

Although further studies are necessary to corroborate preliminary findings, it seems that, in the golden hamster, light adaptation (as indicated by refractoriness of the circadian system to additional light) starts immediately after photic stimulation and reaches a saturation level within 40 minutes. If photic stimulation is shorter than approximately 18 hours, sensitivity is recovered within a day; otherwise, it may take up to 3 weeks.[52] Of course, all of this assumes photic stimulation at maximum-efficiency levels (which, in the hamster, is as low as 100 lux for about a minute). Within the range between threshold and maximum-efficiency levels, stimuli of higher intensity or longer duration will cause greater phase shifts.[55,56]

To summarize the chapter in one sentence, entrainment is the result of daily shifts that correct the difference between the period of the pacemaker and the period of the zeitgeber. More details about the effects of photic stimuli (as well as of non-photic stimuli) on the circadian pacemaker will be discussed in the following chapter.

bibliographic notes

1. Refinetti, R. and Menaker, M. (1992). The circadian rhythm of body temperature. *Physiology and Behavior* 51: 613-637.
2. Konopka, R. J. and Benzer, S. (1971). Clock mutants of *Drosophila melanogaster*. *Proceedings of the National Academy of Sciences* 68: 2112-2116.
3. Feldman, J. F. and Hoyle, M. N. (1973). Isolation of circadian clock mutants of *Neurospora crassa*. *Genetics* 75: 605-613.
4. Ralph, M. R. and Menaker, M. (1989). A mutation of the circadian system in golden hamsters. *Science* 241: 1125-1127.
5. Vitaterna, M. H., King, D. P., Chang, A, M., Kornhauser, J. M., Lowrey, P. L., McDonald, J. D., Dove, W. F., Pinto, L. H., Turek, F. W., and Takahashi, J. S. (1994). Mutagenesis and mapping of a mouse gene, *Clock*, essential for circadian behavior. *Science* 264: 719-725.
6. Refinetti, R. and Menaker, M. (1992). The circadian rhythm of body temperature of normal and tau-mutant golden hamsters. *Journal of Thermal Biology* 17: 129-133.
7. Refinetti, R. and Menaker, M. (1993). Independence of heart rate and circadian period in the golden hamster. *American Journal of Physiology* 264: R235-R238.
8. Refinetti, R. and Menaker, M. (1992). Evidence for separate control of circadian and estrous periodicity in the golden hamster. *Behavioral and Neural Biology* 58: 27-36.
9. Refinetti, R. (1996). Ultradian rhythms of body temperature and locomotor activity in wild-type and tau-mutant hamsters. *Animal Biology* 5: 111-115.
10. Refinetti, R. and Menaker, M. (1997). Is energy expenditure in the hamster primarily under homeostatic or circadian control? *Journal of Physiology* 501: 449-453.
11. Lucas, R. J., Stirland, J. A., Darrow, J. M., Menaker, M., and Loudon, A. S. I. (1999). Free running circadian rhythms of melatonin, luteinizing hormone, and cortisol in Syrian hamsters bearing the circadian *tau* mutation. *Endocrinology* 140: 758-764.
12. Daan, S. and Pittendrigh, C. S. (1976). A functional analysis of circadian pacemakers in nocturnal rodents. II. The variability of phase response curves. *Journal of Comparative Physiology* 106: 253-266.
13. Eskin, A. (1971). Some properties of the system controlling the circadian activity rhythm of sparrows. In: Menaker, M. (Ed.). *Biochronometry*. Washington, DC: National Academy of Sciences, pp. 66-80.
14. Madrid, J. A., Sánchez-Vázquez, F. J., Lax, P., Matas, P., Cuenca, E. M., and Zamora, S. (1998). Feeding behavior and entrainment limits in the circadian system of the rat. *American Journal of Physiology* 275: R372-R383.

15. Davis, F. C. and Menaker, M. (1981). Development of the mouse circadian pacemaker: Independence from environmental cycles. *Journal of Comparative Physiology* 143: 527-539.
16. Refinetti, R. (1998) Influence of early environment on the circadian period of the tau-mutant hamster. *Behavior Genetics* 28: 153-158.
17. Aschoff, J. (1960). Exogenous and endogenous components in circadian rhythms. *Cold Spring Harbor Symposia on Quantitative Biology* 25: 11-28.
18. Richter, C. P. (1971). Inborn nature of the rat's 24-hour clock. *Journal of Comparative and Physiological Psychology* 75: 1-4.
19. Tomioka, K., Uwozumi, K., and Matsumoto, N. (1997). Light cycles given during development affect freerunning period of circadian locomotor rhythm of *period* mutants in *Drosophila melanogaster*. *Journal of Insect Physiology* 43: 297-305.
20. Barrett, R. K. and Page, T. L. (1989). Effects of light on circadian pacemaker development. I. The freerunning period. *Journal of Comparative Physiology A* 165: 41-49.
21. Friesen, W. O., Block, G. D., and Hocker, C. G. (1993). Formal approaches to understanding biological oscillators. *Annual Review of Physiology* 55: 661-681.
22. Pittendrigh, C. S. (1981). Circadian systems: Entrainment. In: Aschoff, J. (Ed.). *Handbook of Behavioral Neurobiology, Volume 4: Biological Rhythms*. New York: Plenum, pp. 95-124.
23. Daan, S. and Pittendrigh, C. S. (1976). A functional analysis of circadian pacemakers in nocturnal rodents. III. Heavy water and constant light: Homeostasis of frequency? *Journal of Comparative Physiology* 106: 267-290.
24. Aschoff, J. (1990). From temperature regulation to rhythm research. *Chronobiology International* 7: 179-186.
25. DeCoursey, P. J. (1960). Daily light sensitivity rhythm in a rodent. *Science* 131: 33-35.
26. Summer, T. L., Ferraro, J. S., and McCormack, C. E. (1984). Phase-response and Aschoff illuminance curves for locomotor activity of the rat. *American Journal of Physiology* 246: R299-R304.
27. Rosenberg, R. S., Zee, P. C., and Turek, F. W. (1991). Phase response curves to light in young and old hamsters. *American Journal of Physiology* 261: R491-R495.
28. Sharma, V. K. (1996). Light-induced phase response curves of the circadian activity rhythm in individual field mice, *Mus booduga*. *Chronobiology International* 13: 401-409.
29. Lee, T. M. and Labyak, S. E. (1997). Free-running rhythms and light- and dark-pulse phase response curves for diurnal *Octodon degus* (Rodentia). *American Journal of Physiology* 273: R278-R286.
30. Kramm, K. R. and Kramm, D. A. (1980). Photoperiodic control of circadian activity rhythms in diurnal rodents. *International Journal of Biometereology* 24: 65-75.
31. Duffield, G. E., Hastings, M. H., and Ebling, F. J. P. (1998). Investigation into the regulation of the circadian system by dopamine and melatonin in the adult Siberian hamster (*Phodopus sungorus*). *Journal of Neuroendocrinology* 10: 871-884.
32. Hoban, T. M. and Sulzman, F. M. (1985). Light effects on circadian timing system of a diurnal primate, the squirrel monkey. *American Journal of Physiology* 249: R274-R280.

33. Drennan, M., Kripke, D. F., and Gillin, J. C. (1989). Bright light can delay human temperature rhythm independent of sleep. *American Journal of Physiology* 257: R136-R141.
34. Shanahan, T. L. and Czeisler, C. A. (1991). Light exposure induces equivalent phase shifts of the endogenous circadian rhythms of circulating plasma melatonin and core body temperature in men. *Journal of Clinical Endocrinology and Metabolism* 73: 227-235.
35. Boivin, D. B., Duffy, J. F., Kronauer, R. E., and Czeisler, C. A. (1996). Dose-response relationships for resetting of human circadian clock by light. *Nature* 379: 540-542.
36. Czeisler, C. A., Kronauer, R. E., Allan, J. S., Duffy, J. F., Jewett, M. E., Brown, E. N., and Ronda, J. M. (1989). Bright light induction of strong (Type 0) resetting of the human circadian pacemaker. *Science* 244: 1328-1333.
37. Honma, K., Honma, S., and Wada, T. (1987). Phase-dependent shift of free-running human circadian rhythms in response to a single bright light pulse. *Experientia* 43: 1205-1207.
38. Czeisler, C. A., Allan, J. S., Strogatz, S. H., Ronda, J. M., Sánchez, R., Ríos, C. D., Freitag, W. O., Richardson, G. S., and Kronauer, R. E. (1986). Bright light resets the human circadian pacemaker independent of the timing of the sleep-wake cycle. *Science* 233: 667-671.
39. Buresová, M., Dvoráková, M., Zvolsky, P., and Illnerová, H. (1991). Early morning bright light phase advances the human circadian pacemaker within one day. *Neuroscience Letters* 121: 47-50.
40. Honma, K. and Honma, S. (1988). A human phase response curve for bright light pulses. *Japanese Journal of Psychiatry and Neurology* 42: 167-168.
41. Dawson, D., Lack, L., and Morris, M. (1993). Phase resetting of the human circadian pacemaker with use of a single pulse of bright light. *Chronobiology International* 10: 94-102.
42. Aschoff, J. and Pohl, H. (1978). Phase relations between a circadian rhythm and its zeitgeber within the range of entrainment. *Naturwissenschaften* 65: 80-84.
43. Wever, R. A. (1989). Light effects on human circadian rhythms: A review of recent Andechs experiments. *Journal of Biological Rhythms* 4: 161-185.
44. Rosenwasser, A. M., Boulos, Z., and Terman, M. (1983). Circadian feeding and drinking rhythms in the rat under complete and skeleton photoperiods. *Physiology and Behavior* 30: 353-359.
45. Refinetti, R., Nelson, D.E., and Menaker, M. (1992). Social stimuli fail to act as entraining agents of circadian rhythms in the golden hamster. *Journal of Comparative Physiology A* 170: 181-187.
46. Honma, K., Honma, S., and Wada, T. (1987). Entrainment of human circadian rhythms by artificial bright light cycles. *Experientia* 43: 572-574.
47. DeCoursey, P. J. (1986). Light-sampling behavior in photoentrainment of a rodent circadian rhythm. *Journal of Comparative Physiology A* 159: 161-169.
48. Pratt, B. L. and Goldman, B. D. (1986). Activity rhythms and photoperiodism of Syrian hamsters in a simulated burrow system. *Physiology and Behavior* 36: 83-89.
49. Daan, S. and Berde, C. (1978). Two coupled oscillators: Simulations of the circadian pacemaker in mammalian activity rhythms. *Journal of Theoretical Biology* 70: 297-313.

50. Refinetti, R. (1993). A functional model of the mammalian circadian pacemaker. *International Journal of Biomedical Computing* 32: 45-60.
51. Shimomura, K. and Menaker, M. (1994). Light-induced phase shifts in *tau* mutant hamsters. *Journal of Biological Rhythms* 9: 97-110.
52. Daymude, J. A. and Refinetti, R. (1999). Phase-shifting effects of single and multiple light pulses in the golden hamster. *Biological Rhythm Research* 30.
53. Sumová, A., Trávníčková, Z., and Illnerová, H. (1995). Memory on long but not on short days is stored in the rat suprachiasmatic nucleus. *Neuroscience Letters* 200: 191-194.
54. Vuillez, P., Jacob, N., Teclemariam-Mesbah, R., and Pévet, P. (1996). In Syrian and European hamsters, the duration of sensitive phase to light of the suprachaismatic nuclei depends on the photoperiod. *Neuroscience Letters* 208: 37-40.
55. Nelson, D. E. and Takahashi, J. S. (1991). Sensitivity and integration in a visual pathway for circadian entrainment in the hamster (*Mesocricetus auratus*). *Journal of Physiology* 439: 115-145.
56. Bauer, M. S. (1992). Irradiance responsivity and unequivocal Type 1 phase responsivity of rat circadian activity rhythms. *American Journal of Physiology* 263: R1110-R1114.

chapter six

---

environmental influences

> The effect of various environmental stimuli (such as
> light, temperature, and food availability) on the circa-
> dian pacemaker is discussed in detail.

In the preceding chapter, we discussed the general mechanism of entrainment of the circadian pacemaker by a zeitgeber. In the present chapter, we
will discuss the role of several environmental stimuli as zeitgebers. A number
of drugs and hormones have also been found to affect the circadian pacemaker but, since they are not — strictly speaking — *environmental* agents,
and since their sites of action in the nervous system are relatively well-
known, they will be dealt with later on in connection with the pharmacology
of the clock (Chapter 9).

photic stimuli

The alternation of day and night is perhaps the most reliable rhythmic
variable in the environment. While seated on the back porch of my house
last night, watching the moonlight reflect on the lake, I measured less than
10 lux of illuminace. Today, in the mid-afternoon with the sun shining behind
me, I am measuring over 5,000 lux. Even on cloudy days, illuminance levels
during the day are several orders of magnitude greater than during the night.
The duration of sunlight does vary from summer to winter, but the variation
is quite regular, and clear differences between day and night are observed
throughout the year except at very extreme latitudes. As a matter of fact, the
duration of sunlight is so reliable that seasonal adaptation in thermoregulatory mechanisms is based mostly on photoperiod rather than on seasonal
variations in ambient temperature.[1-3]

Light has been found to be a robust zeitgeber in a multitude of species,
including those whose visual system is insensitive to light.[4-6] Most humans
in urban areas stay indoors most of the time, but do manage to be exposed
to sunlight a few hours a day. For instance, 10 people in San Diego, California,
were asked to wear photocells on their foreheads continuously, so that levels

of illuminance could be measured. Out of 10 hours of sunlight per day, the subjects were exposed to light an average of 90 minutes.[7] In another study, individuals in San Diego were found to be exposed to sunlight for a little over an hour a day in the winter and a little over 2 hours in the summer, while individuals in Rochester, New York were exposed to about 2.5 hours of sunlight in the summer but less than half an hour in the winter.[8]

As discussed in Chapter 5, entrainment is the result of daily shifts that correct the difference between the period of the pacemaker and the period of the zeitgeber. Consequently, the characteristics of the entrained state depend on three basic elements — the period of the pacemaker, the period of the zeitgeber, and the photic sensitivity of the pacemaker (as described by its phase-response curve). For any given phase-response curve, various characteristics of entrainment can be revealed by manipulation of the period of the pacemaker and the structure of the light-dark cycle. Figure 6 describes eight different situations that have been thoroughly studied in the laboratory.

For clarity of presentation, the actograms shown in Figure 6 were generated by a computer model of the circadian pacemaker of the golden hamster.[9] The top left actogram shows a free-run in constant darkness (often referred to as DD, since constant darkness is a Light-Dark cycle with two Dark phases) of an "animal" with a circadian period of 23.9 hours. It is evident that this animal's period is shorter than 24 hours, as the actogram shows an inclination to the left. On the right, it can be seen that the same animal exhibits a slightly longer period when maintained in constant light (or LL), so that the expressed τ is 24.0 hours even though the true τ is 23.9 hours. This is an instance of Aschoff's rule, according to which τ lengthens as the intensity of light increases in nocturnal animals and shortens as light intensity increases in diurnal animals.[10-17] In actuality, the change in period caused by exposure to constant light is a direct result of the integration of the phase-response curve.[18] Thus, animals with PRCs that have a greater phase-delay than phase-advance region exhibit greater delays (longer periods) in constant light, whereas animals with PRCs that have a greater phase-advance than phase-delay region exhibit greater advances (shorter periods) in constant light. Of course, you may have noticed that the "hamster" in Figure 6 exhibits a longer period in LL than in DD even though its PRC (Figure 5 in Chapter 5) has a greater phase-advance region. The reason for this apparent contradiction is that what really matters is the final integration of discrete phase-shifts, not the algebraic summation of the shifts as read off the static PRC. When the animal is exposed to light at all points of the circadian cycle, the actual response to light is given by a greatly distorted PRC (as each pulse causes a shift, thus altering the circadian time of the following pulse). If it were not for computer simulations, one would never be able to appreciate this difference.

The second row of actograms in Figure 6 shows that a light-dark cycle with a period of 24 hours can easily entrain a pacemaker with a period of 23.9 hours. Although a full light-dark cycle (LD 12:12, which stands for 12 hours of light and 12 hours of darkness) produces entrainment more rapidly

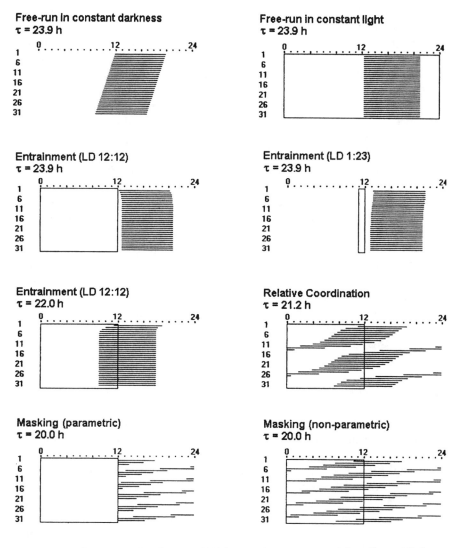

Figure 6. Computer simulations showing various aspects of the effect of light on the circadian system of the golden hamster. In each actogram, the box indicates the duration of the light phase of the light-dark cycle, if applicable.

than daily 1-hour pulses (LD 1:23), both light schedules produce stable entrainment. In the third row, it can be seen that an LD 12:12 cycle entrains a pacemaker with a period of 22 hours but that the *phase angle of entrainment* (that is, the time between lights off and the onset of locomotor activity, symbolized by Ψ) is quite different from that of a pacemaker with a period of 23.9 hours. The difference is easily explained by the phase-shifting requirements: Since a pacemaker with a shorter period needs greater phase delays to match the period of the zeitgeber, more of the phase-delay region of the PRC must be exposed to light (and this is accomplished by moving the circadian system forward into the light phase of the LD cycle). If, however, the period of the pacemaker is too short (such as 21.2 hours in the next actogram), the circadian system cannot shift enough. *Relative coordination* — a clear "attempt" at entrainment without a stable outcome — is observed instead.[19]

In connection with the topic of the phase angle of entrainment, some reference must be made to the *morning–evening* typology. As a result of natural biological variability, some people have larger phase angles of entrainment than others. In 1976, Horne and Östberg created a questionnaire to classify people along a continuum from the morning type (people who usually wake up early and are more productive in the morning) to the evening type (people who usually wake up late and are more productive later in the day).[20] The questionnaire is a valid tool, in the sense that morning types usually wake up around 7:20 am and evening types wake up around 9:20 am.[21] Consistently, the daily rhythm of heart rate peaks earlier (1:30 pm) in morning types than in evening types (5:30 pm).[22] Morning types are more alert at wake time[23] and tend to score higher in measures of achievement tendency and task leadership.[24] They are also better at recognition of sentences presented in the morning than in the afternoon.[25] The morning–evening typology seems to be associated with a genetic polymorphism,[26] although this is difficult to conciliate with the observation that younger people tend to be evening types whereas older people tend to be morning types.[27]

The left actogram in the last row in Figure 6 shows an interesting case that could be interpreted as entrainment by a naive observer. The actogram certainly indicates that the animal is consistently inactive during the day and active during the night, as if it were entrained. However, the computer simulation added a parametric effect of light, namely, the inhibition of activity by light. By examining the actogram to the right, which is identical to the one on the left except for the photic inhibition, one can easily see that the pacemaker is actually free-running through the light-dark cycle. Thus, in the actogram on the left, light simply blocked the expression of the activity rhythm without actually affecting the pacemaker. This blockade, which often is only partial, is called *masking*.[19] Finally, it should be noticed that the apparent free-run in the actogram on the right is not truly a free-run. Close inspection reveals that the period of the activity rhythm becomes slightly longer, and then slightly shorter as the animal enters and traverses the light phase of the LD cycle. This is a natural result of the movement of light stimulation along the different segments of the PRC.

Three well-known effects of light on the circadian system were not described in Figure 6. One of them is the difference in the latency for expression of advances and delays. In real animals maintained in constant darkness, phase advances caused by discrete light pulses often go through several *transients*, in the sense that the activity rhythm advances a little each day until reaching the full phase advance after 4 or 5 days. In contrast, full phase delays can usually be observed on the day immediately following the light pulse. It is still not known why phase advances must go through transients, but the phenomenon clearly lies in the expression of the overt rhythm, not in the state of the pacemaker. That is, the pacemaker itself shifts immediately. This was demonstrated in several ways. For instance, rats maintained under a light-dark cycle were subjected to an 8-hour phase advance of the zeitgeber. It took them, on average, 6.4 days to re-entrain to the new schedule. Rats transferred to constant darkness 2 or 3 days after the shift in the LD cycle showed behavioral shifts equivalent to those eventually shown by rats maintained in the new LD cycle, which indicates that the pacemaker shift had been completed by the time of the transfer.[28] Better yet, mice maintained in constant darkness and subjected to a single light pulse followed by a second pulse later on the same day or on the next day showed behavioral phase shifts consistent with the prediction of additive effects based on the PRC (which would be possible only if the pacemaker had already shifted fully after the first pulse).[29] Similar results were obtained with golden hamsters.[30,31] In addition, it has been reported that, at least in the golden hamster, the difference in shift latency between advances and delays applies only to the onsets of activity — not to the offsets.[32]

Another effect of photic stimulation not yet discussed is the phenomenon of *splitting*. Splitting refers to the separation of a free-running rhythm into two separate components 180° apart. This bizarre phenomenon (see Figure 7) was observed in various animals maintained in constant light for extended periods of time.[33-37] Because the two components exhibit different circadian periods during the initial stage of splitting, some researchers suggested that two separate clocks might be involved. According to this view, the circadian pacemaker would not be a single oscillator, but actually the outcome of the coupling of two or more separate oscillators.[18,38,39] Unlike the two clocks proposed in connection with the phenomenon of internal desynchronization (see Chapter 5), these two oscillators would simply be distinct components of the circadian clock, affecting all physiological variables equally. Nevertheless, good skeptics are never willing to accept multiple elements when the facts can be explained by a single element. The phenomenon of splitting could be seen as an anomaly yet to be explained rather than as a fundamental piece of evidence requiring a revision of the concept of the circadian pacemaker. Indeed, some models of the circadian pacemaker have simply ignored the phenomenon of splitting.[9,40]

For several reasons, I favor the "anomaly" alternative for the explanation of splitting. First, the idea that constant light causes the two putative oscillators to uncouple is inconsistent with the fact that some animals exhibit splitting when maintained in constant darkness or dim light.[41-43] Second, the two split components have identical periods during the split state; the difference in periods occurs only during the very initial stage of splitting and, therefore, could be simply a transient disruption in the expressed rhythms. Third, if the activity rhythm is induced to split and re-unite multiple times, the "evening" and "morning" components trade places, indicating that the two components are not generated by two oscillators with distinct properties.[36] Fourth, although the coupled-oscillators model assumes that the two oscillators respond differently to photic stimulation, empirical investigation of the photic sensitivity of the putative oscillators has failed to demonstrate differential responsivity.[43] Fifth, as discussed in Chapter 8, research on the neural substrate of the circadian pacemaker has not supported the concept of two distinct oscillators. For these reasons, I do not believe that the phenomenon of splitting justifies the assumption of two independent oscillators.

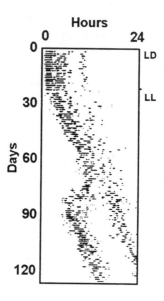

Figure 7.   Actogram showing splitting of the activity rhythm of a golden hamster after prolonged exposure to constant light.

Finally, it must also be pointed out that prolonged exposure to constant light may eliminate the expression of circadian rhythmicity in some species.[12,44-46] The presence of light itself is unlikely to be the important variable, as the loss of rhythmicity has been described in animals maintained in constant darkness as well.[47,48] It would seem that the operation of the circadian clock — or its expression in overt rhythms — requires, in some species, the occasional presence of a zeitgeber.

non-photic stimuli

Several environmental stimuli besides light have been shown to entrain circadian rhythms. Although no class of zeitgebers has been studied in as much detail as photic stimuli, it does seem that the non-parametric theory of entrainment applies to non-photic stimuli as well as it applies to photic stimuli. Based on the data available today, it seems that the phase-response curves to several non-photic stimuli are quite similar to each other but different from the photic phase-response curve. This allows all non-photic PRCs to be grouped together under the label of "non-photic PRC" and be compared with the photic PRC. For instance, while the photic PRC of the golden hamster has a phase-advance peak around CT 19, the non-photic PRC has a phase-advance peak at CT 7 (that is, during subjective day rather than during subjective night).[49] Once again, the existence of two classes of PRCs might be explained by the existence of two separate pacemakers. However, it is much more efficient to assume that *the* pacemaker responds differently to different classes of stimuli.

Given the natural effect of temperature on the speed of biochemical reactions (which is commonly described by a Q_{10} of 2, meaning that the speed doubles for each 10°C increase in temperature), it is natural to expect that the speed of the circadian clock would be affected by changes in environmental temperature — at least in plants and invertebrates, whose body temperature oscillates with the oscillations in the temperature of the environment. This expected parametric effect of temperature, however, was shown not to occur. Research in the 1940s and 1950s revealed that the circadian system of invertebrates is temperature compensated, meaning that it has a biochemical armamentarium to compensate for the potential effect of temperature on the speed of the clock.[50-52] Thus, circadian period is not significantly affected by changes in temperature. Entrainment by temperature cycles must, therefore, be attained by a non-parametric mechanism. That a cycle of ambient temperature can entrain circadian rhythms was shown in plants early in the 20th century.[53] Since then, entrainment by temperature cycles has been demonstrated in invertebrates, birds, mammals, and other animals.[16,54-59] Discrete pulses of temperature change have also been shown to cause phase-shifts of circadian rhythms.[60] In general, it has been found that ambient temperature is a weak zeitgeber, in the sense that not all animals exposed to the cycle actually entrain and, even in those that do, entrainment is often rather unstable.

Social stimuli — that is, stimuli resulting from the interaction of an individual with other individuals of the same species — are a polemical class of zeitgebers. With a few exceptions, carefully conducted studies failed to demonstrate entrainment of circadian rhythms by social stimuli, while less-rigorous studies produced evidence of social entrainment. Of course, the interaction of an animal with another is likely to have a significant impact on the expression of circadian rhythms. The majority of studies that failed to demonstrate actual entrainment (that is, a stimulus-specific modulation of period and phase of circadian rhythms) did identify some form of social masking.[61-70] In rodents, brief exposure ("social pulse") of a male individual to a male intruder (resulting in aggressive episodes) or to a receptive female (resulting in sexual behavior) was found to phase shift the activity rhythm in two studies from one laboratory[71, 72] but not in two other studies from different laboratories.[63,73] Social entrainment was suggested as the explanation for the synchronization of the behavior of bats in a natural cave, but the study did not include appropriate experimental controls.[74] In the degu, the re-entrainment of females (but not males) after a 6-hour phase shift of the LD cycle seemed to be accelerated by the presence of a male or female previously adapted to the new LD cycle.[75] In three species of birds, a cycle of presentation of a species-specific song seemed to produce genuine entrainment.[76,77] In humans studied for 3 weeks in an Arctic field camp, no signs of social entrainment were observed,[70] whereas laboratory studies suggested the occurrence of social entrainment.[78–80] It must be pointed out that, in this case, the field study probably had a more controlled experimental environment than the laboratory studies. For instance, in one of these studies, the only evidence in support of social entrainment was that the members of pairs of subjects remained synchronized for 4 days in the absence of a light-dark cycle,[80] which can hardly be considered sufficient evidence (that is, 4 days is too short an interval to allow the distinction between entrainment and coincident free-runs).

Much better established is the role of physical exercise as a zeitgeber. Free access to running wheels in rodents causes a shortening of the free-running period.[81–83] Pulses of exercise (that is, brief episodes of access to a running wheel or forced treadmill) have been found to consistently phase shift free-running rhythms.[84–91] Also, a daily cycle of exercise produces true entrainment in rats and mice maintained in constant darkness.[84,92,93] As was the case for ambient temperature, exercise seems to be a weaker zeitgeber than light.

The role of feeding (food and water) as a zeitgeber is also very well established. Entrainment of free-running circadian rhythms by the restriction of food or water availability to a few hours each day has been demonstrated in various species.[94–105] Feeding after several days of fasting has also been shown to cause phase shifts, although not necessarily in a phase-dependent manner.[106,107] As in the case of temperature and exercise, feeding was found to be a relatively weak zeitgeber in animals maintained under otherwise constant conditions. However, hunger is obviously a very strong motivational stimulus. Thus, when animals were placed under the conflicting condition of being exposed simultaneously to a light-dark cycle and a cycle of food availability, their behavior showed a compromise between the two drives, and food was often found to be the dominant stimulus.[93–95,108–116]

Curiously, animals under a restricted feeding regimen regularly anticipate the time of feeding — that is, they become active several hours before the food becomes available each day.[94–100] Obviously, this anticipation could be timed by the circadian system. However, animals will anticipate feeding even when they do not entrain to the feeding cycle.[100,109,113,117] This results in a form of "splitting" of two activity components: one of them free-running and the other apparently pseudo-entrained (masked) by the feeding cycle. Because of the long-lasting nature of this form of splitting, and because the feeding component clearly anticipates the availability of food independently of circadian time, a separate timing mechanism must be inferred. Some simple gastrointestinal process being re-initiated daily by the ingestion of food could explain the food-anticipation component. Again, however, there is more to the story: If, after being subjected to restricted feeding for a while, the animals are maintained in constant darkness without any food, the food-anticipation component free-runs.[94,109,113,117] Of course, animals cannot be maintained without food for extended periods of time, but even a free-run of a few days is quite impressive. It strongly suggests that there is an independent "food-entrainable pacemaker." Even more — if the brain site that houses the circadian pacemaker (see Chapter 8) is destroyed, circadian rhythms are eliminated but the ability to anticipate feeding is preserved.[98,99,102,118–123]

So, does the biological clock have a brother? Should we reorganize our vocabulary and speak of two main pacemakers — a "light-entrainable pacemaker" (LEP) and a "food-entrainable pacemaker" (FEP)? Although some researchers have taken this approach, I do not think it is justified. First, the FEP, as its name says, is affected exclusively by food restriction, whereas the true circadian pacemaker is affected by light, temperature, exercise, *and* food restriction. Second, although the FEP can drive a component of the activity rhythm, it clearly does not drive the other component, and there is no indication that it can drive the various other physiological variables that are driven by the circadian pacemaker. Third, despite its apparent ability to free-run, the FEP is strongly dependent on the master pacemaker.[124,125] Fourth, the physical substrate of the FEP has never been identified.[126] For these reasons, I consider the FEP to be a putative pacemaker with a very limited scope of influence. As such, it cannot claim equal standing with the master clock of the body.

bibliographic notes

1. Heldmaier, G., Steinlechner, S., and Rafael, J. (1982). Nonshivering thermo-genesis and cold resistance during seasonal acclimatization in the Djungarian hamster. *Journal of Comparative Physiology* 149: 1-9.
2. Maier, H.A. and Feist, D.D. (1991). Thermoregulation, growth, and reproduction in Alaskan collared lemmings: Role of short day and cold. *American Journal of Physiology* 261: R522-R530.
3. Ellison, G.T.H., Skinner, J.D., and Haim, A. (1992). The relative importance of photoperiod and temperature as cues for seasonal acclimation of thermoregulation in pouched mice (*Saccostomus campestris*: Cricetidae) from southern Africa. *Journal of Comparative Physiology B* 162: 740-746.
4. Foster, R.G., Provencio, I., Hudson, D., Fiske, S., De Grip, W., and Menaker, M. (1991). Circadian photoreception in the retinally degenerate mouse (*rd/rd*). *Journal of Comparative Physiology A* 169: 39-50.
5. Lovegrove, B.G., Heldmeier, G., and Ruf, T. (1993). Circadian activity rhythms in colonies of "blind" molerats, *Cryptomys damarensis* (Bathyergidae). *South African Journal of Zoology* 28: 46-55.

6. Lovegrove, B.G. and Papenfus, M.E. (1995). Circadian activity rhythms in the solitary cape mole rat (*Georychus capensis*: Bathyergidae) with some evidence of splitting. *Physiology and Behavior* 58: 679-685.
7. Savides, T.S., Messin, S., Senger, C., and Kripke, D.F. (1986). Natural light exposure of young adults. *Physiology and Behavior* 38: 571-574.
8. Cole, R.J., Kripke, D.F., Wisbey, J., Mason, W.J., Gruen, W., Hauri, P.J., and Juarez, S. (1995). Seasonal variation in human illumination exposure at two different latitudes. *Journal of Biological Rhythms* 10: 324-334.
9. Refinetti, R. (1993). A functional model of the mammalian circadian pacemaker. *International Journal of Biomedical Computing* 32: 45-60.
10. Aschoff, J. (1979). Circadian rhythms: Influences of internal and external factors on the period measured in constant conditions. *Zeitschrift für Tierpsychologie* 49: 225-249.
11. Fuller, C.A. and Edgar, D.M. (1986). Effects of light intensity on the circadian temperature and feeding rhythms in the squirrel monkey. *Physiology and Behavior* 36: 687-691.
12. Rosenwasser, A.M. (1993). Circadian drinking rhythms in SHR and WKY rats: Effects of increasing light intensity. *Physiology and Behavior* 53: 1035-1041.
13. Witting, W., Boerma, D., Koster Van Hoffen, G.C., Swaab, D.F., and Mirmiran, M. (1995). Light suppresses frequency and endogenous amplitude of the circadian system in nocturnal animals. *Biological Rhythm Research* 26: 477-485.
14. Lee, T.M. and Labyak, S.E. (1997). Free-running rhythms and light- and dark-pulse phase response curves for diurnal *Octodon degus* (Rodentia). *American Journal of Physiology* 273: R278-R286.
15. Summer, T.L., Ferraro, J.S., and McCormack, C.E. (1984). Phase-response and Aschoff illuminance curves for locomotor activity of the rat. *American Journal of Physiology* 246: R299-R304.
16. Tokura, H. and Aschoff, J. (1983). Effects of temperature on the circadian rhythm of pig-tailed macaques *Macaca nemestrina*. *American Journal of Physiology* 245: R800-R804.
17. Warren, W.S. and Cassone, V.M. (1995). The pineal gland: Photoreception and coupling of behavioral, metabolic, and cardiovascular circadian outputs. *Journal of Biological Rhythms* 10: 64-79.
18. Daan, S. and Pittendrigh, C.S. (1976). A functional analysis of circadian pacemakers in nocturnal rodents. III. Heavy water and constant light: Homeostasis of frequency? *Journal of Comparative Physiology* 106: 267-290.
19. Aschoff, J. (1981). Freerunning and entrained circadian rhythms. In: Aschoff, J. (Ed.). *Handbook of Behavioral Neurobiology, Volume 4: Biological Rhythms*. New York: Plenum, pp. 81-93.
20. Horne, J.A. and Östberg, O. (1976). A self-assessment questionnaire to determine Morningness-Eveningness. *International Journal of Chronobiology* 4: 97-110.
21. Horne, J.A. and Östberg, O. (1977). Individual differences in human circadian rhythms. *Biological Psychology* 5: 179-190.
22. Taillard, J., Sanchez, P., Lemoine, P., and Mouret, J. (1990). Heart rate circadian rhythm as a biological marker of desynchronization in major depression: A methodological and preliminary report. *Chronobiology International* 7: 305-316.
23. Carrier, J., Monk, T.H., Buysse, D.J., and Kupfer, D.J. (1997). Sleep and morning-eveningness in the "middle" years of life (20-59 y). *Journal of Sleep Research* 6: 230-237.

24. Watts, B.L. (1982). Individual differences in circadian activity rhythms and their effects on room mate relationships. *Journal of Personality* 50: 374-384.

25. Natale, V. and Lorenzetti, R. (1997). Influences of morningness-eveningness and time of day on narrative comprehension. *Personality and Individual Differences* 23: 685-690.

26. Katzenberg, D., Young, T., Finn, L., Lin, L., King, D.P., Takahashi, J.S., and Mignot, E. (1998). A CLOCK polymorphism associated with human diurnal preference. *Sleep* 21: 569-576.

27. May, C.P., Hasher, L., and Stoltfus, E.R. (1993). Optimal time of day and the magnitude of age differences in memory. *Psychological Science* 4: 326-330.

28. Takamure, M., Murakami, N., Takahashi, K., Kuroda, H., and Etoh, T. (1991). Rapid reentrainment of the circadian clock itself, but not the measurable activity rhythms, to a new light-dark cycle in the rat. *Physiology and Behavior* 50: 443-449.

29. Sharma, V.K. and Chandrashekaran, M.K. (1997). Rapid phase resetting of a mammalian circadian rhythm by brief light pulses. *Chronobiology International* 14: 537-548.

30. Daymude, J.A. and Refinetti, R. (1999). Phase-shifting effects of single and multiple light pulses in the golden hamster. *Biological Rhythm Research* 30.

31. Best, J.D., Maywood, E.S., Smith, K.L., and Hastings, M.H. (1999). Rapid resetting of the mammalian circadian clock. *Journal of Neuroscience* 19: 828-835.

32. Meijer, J.H. and De Vries, M.J. (1995). Light-induced phase shifts in onset and offset of running-wheel activity in the Syrian hamster. *Journal of Biological Rhythms* 10: 4-16.

33. Pittendrigh, C.S. and Daan, S. (1976). A functional analysis of circadian pacemakers in nocturnal rodents. V. Pacemaker structure: A clock for all seasons. *Journal of Comparative Physiology* 106: 333-355.

34. Boulos, Z. and Terman, M. (1979). Splitting of circadian rhythms in the rat. *Journal of Comparative Physiology* 134: 75-83.

35. Swan, J.M. and Turek, F.W. (1985). Multiple circadian oscillators regulate the timing of behavioral and endocrine rhythms in female golden hamsters. *Science* 228: 898-900.

36. Earnest, D.J. and Turek, F.W. (1982). Splitting of the circadian rhythm of activity in hamsters: Effects of exposure to constant darkness and subsequent re-exposure to constant light. *Journal of Comparative Physiology* 145: 405-411.

37. Pickard, G.E., Kahn, R., and Silver, R. (1984). Splitting of the circadian rhythm of body temperature in the golden hamster. *Physiology and Behavior* 32: 763-766.

38. Daan, S. and Berde, C. (1978). Two coupled oscillators: Simulations of the circadian pacemaker in mammalian activity rhythms. *Journal of Theoretical Biology* 70: 297-313.

39. D'ez-Noguera, A. (1994). A functional model of the circadian system based on the degree of intercommunication in a complex system. *American Journal of Physiology* 267: R1118-R1135.

40. Lloyd, A.L. and Lloyd, D. (1993). Hypothesis: The central oscillator of the circadian clock is a controlled chaotic attractor. *BioSystems* 29: 77-85.

41. Underwood, H. and Edmonds, K. (1995). The circadian rhythm of thermoregulation in Japanese quail. II. Multioscillator control. *Journal of Biological Rhythms* 10: 234-247.

42. Lovegrove, B.G. and Papenfus, M.E. (1995). Circadian activity rhythms in the solitary cape mole rat (*Georychus capensis*: Bathyergidae) with some evidence of splitting. *Physiology and Behavior* 58: 679-685.
43. Meijer, J.H., Daan, S., Overkamp, G.J.F., and Hermann, P.M. (1990). The two-oscillator circadian system of tree shrews (*Tupaia belangeri*) and its response to light and dark cycles. *Journal of Biological Rhythms* 5: 1-16.
44. Deprés-Brummer, P., Metzger, G., and Lévi, F. (1996). Analyses des rythmes dela température corporelle du rat en libre cours. *Pathologie Biologie* 44: 150-156.
45. Honma, K. and Hiroshige, T. (1978). Simultaneous determination of circadian rhythms of locomotor activity and body temperature in the rat. *Japanese Journal of Physiology* 28: 159-169.
46. Eastman, C. and Rechtschaffen, A. (1983). Circadian temperature and wake rhythms of rats exposed to prolonged continuous illumination. *Physiology and Behavior* 31: 417-427.
47. Kuwabara, N., Seki, K., and Aoki, K. (1986). Circadian, sleep and brain temperature rhythms in cats under sustained daily light-dark cycles and constant darkness. *Physiology and Behavior* 38: 283-289.
48. Gerkema, M.P., Groos, G.A., and Daan, S. (1990). Differential elimination of circadian and ultradian rhythmicity by hypothalamic lesions in the common vole, *Microtus arvalis*. *Journal of Biological Rhythms* 5: 81-95.
49. Smith, R.D., Turek, F.W., and Takahashi, J.S. (1992). Two families of phase-response curves characterize the resetting of the hamster circadian clock. *American Journal of Physiology* 262: R1149-R1153.
50. Brown, F.A. and Webb, H.M. (1948). Temperature relations of an endogenous daily rhythmicity in the fiddler crab, *Uca*. *Physiological Zoology* 21: 371-381.
51. Pittendrigh, C.S. (1954). On temperature independence in the clock system controlling emergence time in *Drosophila*. *Proceedings of the National Academy of Sciences* 40: 1018-1029.
52. Hoffmann, K. (1957). Über den Einfluß der Temperatur auf die Tagesperiodik bei einem Poikilothermen. *Naturwissenschaften* 44: 358.
53. Stern, K. and Bünning, E. (1929).Über die tagesperiodischen Bewegungen der Primärblätter von *Phaseolus multiflorus*. I. Der Einfluß der Temperatur auf die Bewegungen. *Berichte der Deutschen Botanische Gesellschaft* 47: 565-584.
54. North, R.D. (1993). Entrainment of the circadian rhythm of locomotor activity in wood ants by temperature. *Animal Behaviour* 45: 393-397.
55. Eskin, A. (1971). Some properties of the system controlling the circadian activity rhythm of sparrows. In: Menaker, M. (Ed.). *Biochronometry.* Washington, DC: National Academy of Sciences, pp. 66-80.
56. Lindberg, R.G. and Hayden, P. (1974). Thermoperiodic entrainment of arousal from torpor in the little pocket mouse, *Perognathus longimembris*. *Chronobiologia* 1: 356-361.
57. Francis, A.J.P. and Coleman, G.J. (1988). The effect of ambient temperature cycles upon circadian running and drinking activity in male and female laboratory rats. *Physiology and Behavior* 43: 471-477.
58. Rajaratnam, S.M.W. and Redman, J.R. (1998). Entrainment of activity rhythms to temperature cycles in diurnal palm squirrels. *Physiology and Behavior* 63: 271-277.

59. Pohl, H. (1998). Temperature cycles as zeitgebers for the circadian clock of two burrowing rodents, the normothermic antelope ground squirrel and the heterothermic Syrian hamster. *Biological Rhythm Research* 29: 311-325.

60. Francis, A.J.P. and Coleman, G.J. (1997). Phase response curves to ambient temperature pulses in rats. *Physiology and Behavior* 62: 1211-1217.

61. Richter, C.P. (1970). Dependence of successful mating in rats on functioning of the 24-hour clocks of the male and female. *Communications in Behavioral Biology* 5: 1-5.

62. Crowley, M. and Bovet, J. (1980). Social synchronization of circadian rhythms in deer mice (Peromyscus maniculatus). *Behavioral Ecology and Sociobiology* 7: 99-105.

63. Refinetti, R., Nelson, D.E., and Menaker, M. (1992). Social stimuli fail to act as entraining agents of circadian rhythms in the golden hamster. *Journal of Comparative Physiology A* 170: 181-187.

64. Gattermann, R. and Weinandy, R. (1997). Lack of social entrainment of circadian activity rhythms in the solitary golden hamster and in the highly social Mongolian gerbil. *Biological Rhythm Research* 28: 85-93.

65. Goel, N. and Lee, T.M. (1997). Social cues modulate free-running circadian activity rhythms in the diurnal rodent, *Octodon degus*. *American Journal of Physiology* 273: R797-R804.

66. Aschoff, J. and von Goetz, C. (1988). Masking of circadian activity rhythms in male golden hamsters by the presence of females. *Behavioral Ecology and Sociobiology* 22: 409-412.

67. Kleinknecht, S. (1985). Lack of social entrainment of free-running circadian activity rhythms in the Australian sugar glider (*Petaurus breviceps*: Marsupialia). *Behavioral Ecology and Sociobiology* 16: 189-193.

68. Erkert, H.G., Nagel, B., and Stephani, I. (1986). Light and social effects on the free-running circadian activity rhythm in common marmosets (*Callithrix jacchus*: Primates): Social masking, pseudo-splitting, and relative coordination. *Behavioral Ecology and Sociobiology* 18: 443-452.

69. Büttner, D. (1992). Social influences on the circadian rhythm of locomotor activity and food intake of guinea pigs. *Journal of Interdisciplinary Cycle Research* 23: 100-112.

70. Steel, G.D., Callaway, M., Suedfeld, P., and Palinkas, L. (1995). Human sleep-wake cycles in the high Arctic: Effects of unusual photoperiodicity in a natural setting. *Biological Rhythm Research* 26: 582-592.

71. Mrosovsky, N. (1988). Phase response curves for social entrainment. *Journal of Comparative Physiology A* 162: 35-46.

72. Honrado, G.I. and Mrosovsky, N. (1989). Arousal by sexual stimuli accelerates the re-entrainment of hamsters to phase advanced light-dark cycles. *Behavioral Ecology and Sociobiology* 25: 57-63.

73. Meerlo, P. and Daan, S. (1998). Aggressive and sexual social stimuli do not phase shift the circadian temperature rhythm in rats. *Chronobiology International* 15: 231-240.

74. Marimuthu, G., Rajan, S., and Chandrashekaran, M.K. (1981). Social entrainment of the circadian rhythm in the flight activity of the microchiropteran bat *Hipposideros speoris*. *Behavioral Ecology and Sociobiology* 8: 147-150.

75. Goel, N. and Lee, T.M. (1995). Social cues accelerate reentrainment of circadian rhythms in diurnal female *Octodon degus* (Rodentia: Octodontidae). *Chronobiology International* 12: 311-323.

76. Gwinner, E. (1966). Entrainment of a circadian rhythm in birds by species-specific song cycles. *Experientia* 22: 765-766.
77. Menaker, M. and Eskin, A. (1966). Entrainment of circadian rhythms by sound in *Passer domesticus*. *Science* 154: 1579-1581.
78. Wever, R. A. (1989). Light effects on human circadian rhythms: A review of recent Andechs experiments. *Journal of Biological Rhythms* 4: 161-185.
79. Wever, R. (1970). Zur Zeitgeber-Stircadiane Periodik des Menschen. *PflŸgers Archiv* 321: 133-142.
80. Aschoff, J., Fatranská, M., Giedke, H., Doerr, P., Stamm, D., and Wisser, H. (1971). Human circadian rhythms in continuous darkness: Entrainment by social cues. *Science* 171: 213-215.
81. Yamada, N., Shimoda, K., Ohi, K., Takahashi, S., and Takahashi, K. (1988). Free-access to a running wheel shortens the period of free-running rhythm in blinded rat. *Physiology and Behavior* 42: 87-91.
82. Edgar, D.M., Martin, C.E., and Dement, W.C. (1991). Activity feedback to the mammalian circadian pacemaker: Influence on observed measures of rhythm period length. *Journal of Biological Rhythms* 6: 185-199.
83. Edgar, D.M., Kilduff, T.S., Martin, C.E., and Dement, W.C. (1991). Influence of running wheel activity on free-running sleep/wake and drinking circadian rhythms in mice. *Physiology and Behavior* 50: 373-378.
84. Reebs, S.G. and Mrosovsky, N. (1989). Effects of induced wheel running on the circadian activity rhythms of Syrian hamsters: Entrainment and phase response curve. *Journal of Biological Rhythms* 4: 39-48.
85. Mrosovsky, N., Salmon, P.A., Menaker, M., and Ralph, M.R. (1992). Nonphotic phase shifting in hamster clock mutants. *Journal of Biological Rhythms* 7: 41-49.
86. Wickland, C.R. and Turek, F.W. (1991). Phase-shifting effects of acute increases in activity on circadian locomotor rhythms in hamsters. *American Journal of Physiology* 261: R1109-R1117.
87. Janik, D. and Mrosovsky, N. (1993). Nonphotically induced phase shifts of circadian rhythms in the golden hamster: Activity-response curves at different ambient temperatures. *Physiology and Behavior* 53: 431-436.
88. Mistlberger, R.E., Sinclair, S.V., Marchant, E.G., and Neil, L. (1997). Phase shifts to refeeding in the Syrian hamster mediated by running activity. *Physiology and Behavior* 61: 273-278.
89. Reebs, S.G. and Doucet, P. (1997). Relationship between circadian period and size of phase shifts in Syrian hamsters. *Physiology and Behavior* 61: 661-666.
90. Gannon, R.L. and Rea, M.A. (1995). Twelve-hour phase shifts of hamster circadian rhythms elicited by voluntary wheel running. *Journal of Biological Rhythms* 10: 196-210.
91. Sinclair, S.V. and Mistlberger, R.E. (1997). Scheduled activity reorganizes circadian phase of Syrian hamsters under full and skeleton photoperiods. *Behavioural Brain Research* 87: 127-137.
92. Edgar, D.M and Dement, W.C. (1991). Regularly scheduled voluntary exercise synchronizes the mouse circadian clock. *American Journal of Physiology* 261: R928-R933.
93. Mistlberger, R.E. (1991). Effects of daily schedules of forced activity on free-running rhythms in the rat. *Journal of Biological Rhythms* 6: 71-80.
94. Edmonds, S. C. and Adler, N. T. (1977). Food and light as entrainers of circadian running activity in the rat. *Physiology and Behavior* 18: 915-919.

95. Sánchez-Vázquez, F.J., Madrid, J.A., Zamora, S. and Tabata, M. (1997). Feeding entrainment of locomotor activity rhythms in the goldfish is mediated by a feeding-entrainable circadian oscillator. *Journal of Comparative Physiology A* 181: 121-132.

96. Jilge, B. (1991). Restricted feeding: A nonphotic zeitgeber in the rabbit. *Physiology and Behavior* 51: 157-166.

97. Hau, M. and Gwinner, E. (1992). Circadian entrainment by feeding cycles in house sparrow, *Passer domesticus. Journal of Comparative Physiology A* 170: 403-409.

98. Abe, H. and Rusak, B. (1992). Anticipatory activity and entrainment of circadian rhythms in Syrian hamsters exposed to restricted palatable diets. *American Journal of Physiology* 263: R116-R124.

99. Mistlberger, R.E. (1993). Circadian properties of anticipatory activity to restricted water access in suprachiasmatic-ablated hamsters. *American Journal of Physiology* 264: R22-R29.

100. Kennedy, G.A., Coleman, G.J., and Armstrong, S.M. (1995). Entrainment of circadian wheel-running rhythms of the northern brown bandicoot, *Isoodon macrourus,* by daily restricted feeding schedules. *Chronobiology International* 12: 176-187.

101. Challet, E., Malan, A., and Prains circadian rhythms of wheel-running and body temperature in rats kept in constant darkness. *Neuroscience Letters* 211: 1-4.

102. Rusak, B., Mistlberger, R.E., Losier, B., and Jones, C.H. (1988). Daily hoarding opportunity entrains the pacemaker for hamster activity rhythms. *Journal of Comparative Physiology A* 164: 165-171.

103. Sulzman, F.M., Fuller, C.A., and Moore-Ede, M.C. (1977). Environmental synchronizers of squirrel monkey circadian rhythms. *Journal of Applied Physiology* 43: 795-800.

104. Boulos, Z., Frim, D.M., Dewey, L.K., and Moore-Ede, M.C. (1989). Effects of restricted feeding schedules on circadian organization in squirrel monkeys. *Physiology and Behavior* 45: 507-515.

105. Sulzman, F.M., Fuller, C.A., and Moore-Ede, M.C. (1977). Feeding time synchronizes primate circadian rhythms. *Physiology and Behavior* 18: 775-779.

106. Coleman, G.J .and Francis, A.J.P. (1991). Food deprivation and reinstatement phase shifts rat activity rhythms in constant light but not constant dark. *Physiology and Behavior* 50: 167-171.

107. Challet, E., Pévet, P., and Malan, A. (1997). Effect of prolonged fasting and subsequent refeeding on free-running rhythms of temperature and locomotor activity in rats. *Behavioural Brain Research* 84: 275-284.

108. Hotz, M.M., Connolly, M.S., and Lynch, C.B. (1987). Adaptation to daily meal-timing and its effect on circadian temperature rhythms in two inbred strains of mice. *Behavior Genetics* 17: 37-51.

109. Kennedy, G.A., Coleman, G.J., and Armstrong, S.M. (1991). Restricted feeding entrains circadian wheel-running activity rhythms in the kowari. *American Journal of Physiology* 261: R819-R827.

110. Jilge, B. and Stähle: Impact of conflicting zeitgebers on circadian rhythms of the rabbit. *American Journal of Physiology* 264: R708-R715.

111. Rashotte, M.E. and Stephan, F.K. (1996). Coupling between light- and food-entrainable circadian oscillators in pigeons. *Physiology and Behavior* 59: 1005-1010.

112. Hau, M. and Gwinner, E. (1997). Adjustment of house sparrow circadian rhythms to a simultaneously applied light and food zeitgeber. *Physiology and Behavior* 62: 973-981.

113. Honma, K., von Goetz, C., and Aschoff, J. (1983). Effects of restricted daily feeding on freerunning circadian rhythms in rats. *Physiology and Behavior* 30: 905-913.

114. Krieger, D.T. (1974). Food and water restriction shifts corticosterone, temperature, activity and brain amine periodicity. *Endocrinology* 95: 1195-1201.

115. Nelson, W., Scheving, L., and Halberg, F. (1975). Circadian rhythms in mice fed a single daily meal at different stages of lighting regimen. *Journal of Nutrition* 105: 171-184.

116. Mistlberger, R.E. (1991). Scheduled daily exercise or feeding alters the phase of photic entrainment in Syrian hamsters. *Physiology and Behavior* 50: 1257-1260.

117. Phillips, D.L., Rautenberg, W., Rashotte, M.E., and Stephan, F.K. (1993). Evidence for a separate food-entrainable circadian oscillator in the pigeon. *Physiology and Behavior* 53: 1105-1113.

118. Stephan, F.K. (1992). Resetting of a feeding-entrainable circadian clock in the rat. *Physiology and Behavior* 52: 985-995.

119. Mistlberger, R.E. (1992). Nonphotic entrainment of circadian activity rhythms in suprachiasmatic nuclei-ablated hamsters. *Behavioral Neuroscience* 106: 192-202.

120. Krieger, D.T., Hauser, H., and Krey, L.C. (1977). Suprachiasmatic nuclear lesions do not abolish food-shifted circadian adrenal and temperature rhythmicity. *Science* 197: 398-399.

121. Stephan, F.K., Swann, J.M., and Sisk, C.L. (1979). Entrainment of circadian rhythms by feeding schedules in rats with suprachiasmatic lesions. *Behavioral and Neural Biology* 25: 545-554.

122. Clarke, J.D. and Coleman, G.J. (1986). Persistent meal-associated rhythms in SCN-lesioned rats. *Physiology and Behavior* 36: 105-113.

123. Mistlberger, R.E., de Groot, M.H.M., Bossert, J.M., and Marchant, E.G. (1996). Discrimination of circadian phase in intact and suprachiasmatic nuclei-ablated rats. *Brain Research* 739: 12-18.

124. Stephan, F.K. (1986). Coupling between feeding- and light-entrainable circadian pacemakers in the rat. *Physiology and Behavior* 38: 537-544.

125. Ottenweller, J.E., Tapp, W.N., and Natelson, B.H. (1990). Phase-shifting the light-dark cycle resets the food-entrainable circadian pacemaker. *American Journal of Physiology* 258: R994-R1000.

126. Mistlberger, R.E. and Rusak, B. (1988). Food-anticipatory circadian rhythms in rats with paraventricular and lateral hypothalamic ablations. *Journal of Biological Rhythms* 3: 277-291.

chapter seven

complex parameters

Parameters of circadian rhythms that are not immediately explained by the operation of the circadian clock (such as waveform and amplitude of oscillation) are examined.

In the three preceding chapters, we saw that circadian rhythms are generated by an endogenous clock — the circadian pacemaker — and that this clock can be entrained by environmental stimuli. Therefore, we addressed two main parameters of circadian rhythms: their free-running period (τ) and their phase relationship with the zeitgeber (Ψ). However, rhythmic processes are not fully characterized by only their period and phase. Three other parameters are necessary for a full characterization: the amplitude (excursion), the mean level, and the waveform of the oscillation. Indeed, consider the circadian rhythm of body temperature. To say that body temperature oscillates with a period of about 25 hours in the absence of external time cues, and that it peaks at about 5 pm in individuals living under standard zeitgeber conditions, does not tell the whole story. How much does temperature oscillate each day? Does it oscillate around a mean level of 0°C or of 37°C? Does it jump up in the morning and down in the evening, or does it oscillate smoothly? Although the amplitude, mean level, and waveform of circadian rhythms are presumably determined to some extent by the circadian clock, they are greatly affected by the peculiarities of the various effector mechanisms — and, therefore, they are much more complex than period and phase. Due to the greater complexity, much less is known about their properties.

Although there is no standard terminology adopted by all researchers of biological rhythms, some commonly used terms must be presented at this point. The period of circadian rhythms is usually called just that — *period* — and it is expressed in hours. In constant darkness, the period of a circadian rhythm is assumed to be a direct reflection of the period of the circadian pacemaker. The inverse of period — frequency — is often used for various other biological rhythms. Thus, one says that alpha waves in the electroencephalogram have a frequency of about 10 Hz, which is the same as saying that the period of an alpha wave is 0.1 second. On the other hand, one never says that the frequency of circadian rhythms is about 11.6 µHz, even though it would be technically correct to do so.

Researchers describing circadian rhythms sometimes divide the duration of a circadian cycle (as determined by its period) into two segments: the segment during which the animals are active (denoted by the Greek lower-case letter *alpha*, α) and the segment during which the animals are resting (denoted by *rho*, ρ). In nocturnal animals, α occurs during the night, while in diurnal animals it occurs during the day.

When referring to the phase relationship between two rhythms (or between a rhythm and a zeitgeber), one needs to determine a point of reference. For activity rhythms plotted on an actogram, the onset of activity is usually the chosen point of reference. For rhythms plotted as true waves, the time of the daily peak (or trough) is usually employed. The time of the daily peak of a circadian rhythm is called the *acrophase*. Phase differences between the acrophases of two rhythms (or between the acrophase of a rhythm and an environmental event) can be expressed either in hours or in degrees (where 360°corresponds to a full cycle).

The amplitude of the rhythm is usually called just that — *amplitude* — although some authors refer to it as the *range of oscillation*. Researchers with a physics background usually define the amplitude as half the excursion of the oscillation (which is standard practice in engineering). Researchers with a biology background usually refer to the amplitude as the full excursion of the oscillation (the difference between *peak* and *trough* — or the difference between *zenith* and *nadir*, to use the astronomer's terminology).

The mean level around which the variable oscillates is often — although certainly not always — referred to as the *mesor* of the rhythm. For a sym-metrical oscillation, it corresponds to the midpoint between the peak and the trough. Otherwise, it is usually calculated as the arithmetic mean of all values in a cycle (or many successive cycles).

Due to biological variability, the *waveform* of circadian rhythms is never constant. Besides the variability in ultradian oscillations superimposed on the circadian oscillation, the shape of the circadian component itself varies from day to day. In some cases, the waveform can be approximated by a simple cosine wave.[1] Usually, however, circadian rhythms are rather irregu-lar and often exhibit two peaks in each cycle.[2] Non-parametric procedures that do not require assumptions about the regularity of the rhythm are, therefore, necessary for the analysis of circadian rhythms.[3-5] The waveform of circadian rhythms is expected to vary depending on the proportion of light and darkness in the light-dark cycle, as most functions (such as loco-motor activity, feeding, drinking, and mating) are usually restricted to either the light phase (in diurnal animals) or the dark phase (in nocturnal animals). In the only systematic study conducted so far, pigeons were indeed found to compress the active phase of their rhythms as the duration of the light portion of the light-dark cycle was reduced from 21 hours per day to 3 hours per day.[6]

It is rather obvious that different circadian rhythms in the same organism have different mean levels and amplitudes. After all, the units of measurement themselves are different. For instance, in the golden hamster, body temperature oscillates from 36.2 to 37.0°C with a mean of 36.6°C,[7] heart rate oscillates from 370 to 460 bpm with a mean of 400 bpm,[8] and melatonin content in the pineal gland oscillates from 0 to 700 pg with a mean of 90 pg.[9] As oranges and apples cannot be compared, so the mean levels and amplitudes of different rhythms cannot be compared. Of course, phase relationships are a different story, since the acrophases of different rhythms can be easily compared, as long as the waveforms are similar. Different rhythms in the same animal do have different acrophases and, if one measures a large enough number of variables, one can find at least one rhythm peaking at any given time of the circadian cycle.[10]

While mean levels and amplitudes cannot be compared across variables, they can certainly be compared across species within the same variable. Regarding the body temperature rhythm, there are considerable interspecies differences in amplitude, as exemplified in Figure 3 (Chapter 4). As seen there, the amplitude of the daily oscillation is less than 2°C in the rat but more than 4°C in the tree shrew. Although the robustness (regularity) of the rhythm seems to be greater in larger animals, the amplitude does not show a regular pattern across species.[11] The mean level is usually in the range of 36.5 to 38.5°C in mammals and of 40 to 42°C in birds, although noneutherian mammals and very small birds often have much lower levels of body temperature.[12-14]

By far the best studied rhythms have been those of locomotor activity and body temperature. As exemplified in Figure 8, the temporal course of the two rhythms is very similar in the ground squirrel. Body temperature rises when the animal becomes more active and falls when activity is reduced, even though individual bouts of activity are not necessarily correlated with individual rises in body temperature. In contrast, the human body temperature rhythm has been consistently found to start to ascend several hours before awakening.[15-18] The relationship between the two rhythms has rarely been analyzed systematically in other animals, but two studies of mine, one in rats[19] and another in eight different mammalian species,[20] provided results equivalent to those shown in Figure 8 (that is, an almost perfect synchrony between the two rhythms). This suggests that humans and rodents differ in the phase relationship between the rhythms of activity and body temperature. In actuality, the difference is only quantitative. In both humans and rodents, temperature starts to ascend slowly several hours before awakening (more so in humans than in squirrels) and then rises abruptly (more so in rodents than in humans) at wake time.[20]

With or without a delay between them, the rhythms of body temperature and activity proceed so closely together that it is quite reasonable to wonder whether the temperature rhythm is not simply a consequence of the activity

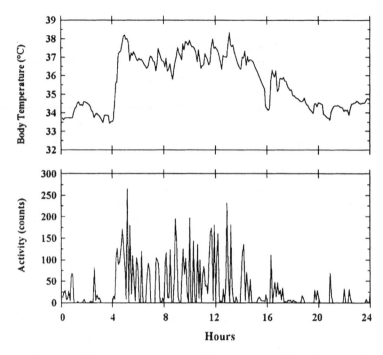

Figure 8.   Records of body temperature and locomotor activity of a ground squirrel recorded every 6 minutes for 24 hours.

rhythm. Indeed, disruption of the body temperature rhythm by changes in activity was observed as early as 1884,[21] and the fact that acute episodes of physical activity and exercise can elevate body temperature has been extensively documented in humans[22–25] and other vertebrates.[26–29] Consequently, the daily elevation in body temperature might very well be a direct result of increased activity. In order to investigate the question, several researchers recorded the body temperature rhythm of human subjects maintained in continuous bed rest. Although the amplitude of the rhythm was reduced under this condition, robust rhythmicity persisted.[30–33] For instance, in a study where the subjects were maintained in bed rest for 6 consecutive days, only a small reduction in the amplitude of the temperature rhythm (from 1.5°C to 1.3°C) was observed.[33] Thus, while the activity rhythm may alter the amplitude and shape of the body temperature rhythm, it does not cause it. The same was shown to be true in several small mammals by a comparison of diurnal and nocturnal body temperatures associated with the same levels of activity.[20,34–36]

If the body temperature rhythm is not a side effect of the activity rhythm, it must be generated independently by the circadian clock. If so, the signal from the clock should be present even when effector mechanisms are not available. That is, animals that do not possess the metabolic machinery to elevate body temperature above the temperature of the environment (the so-called *ectotherms*) should nevertheless have the "motivation" to have a circadian rhythm of body temperature. When placed in a thermal gradient that allows them to select the temperature of their environment, they should display a rhythmic pattern of selection. Indeed, crustaceans, fishes, and lizards were shown to display a daily rhythm of temperature selection — which resulted in a daily rhythm of body temperature — not only under a light–dark cycle[37-41] but under constant conditions as well.[41-43] Large iguana lizards, which can retain a small amount of metabolic heat, were found to exhibit a circadian rhythm of body temperature even in the absence of a thermal gradient.[44]

When *endotherms* were tested in a thermal gradient, curious results were obtained. Rats and hamsters did exhibit a daily rhythm of temperature selection, but the rhythm was 180°out of phase with their endogenous body temperature rhythm.[40-48] That is, the animals selected higher ambient temperatures when their body temperature was low, and vice versa. This suggests that the thermal motivation was directed at combating the oscillation in body temperature imposed by the circadian system. "Thermoregulatory system declares war against the circadian system" — could have been the headline in the body's newspaper.

A conflict between homeostatic and circadian mechanisms is a rather disturbing idea and, consequently, I embarked on an extensive research project to elucidate the matter. The project was supported by the National Science Foundation and the National Institutes of Health, which I gratefully acknowledge. First, I replicated the findings in the golden hamster to ensure that the results were not an artifact of the methods utilized in a specific laboratory. To allow the animals to select the temperature of their environment, I housed them (one at a time) inside a long cage placed inside a metal tube that was heated on one side and cooled on the other, thus creating an air-temperature gradient ranging from 14°C (57°F) to 33°C (91°F). Food and water were available at various points along the cage. Rather than sitting down and watching the hamsters 24 hours a day, I used a multiple infrared-photocell system and a computer to continuously monitor the position of the animal along the temperature gradient. Body temperature was also recorded continuously by telemetry (using a tiny temperature-sensitive radio transmitter previously implanted in the abdominal cavity under general anesthesia). The animals were left undisturbed for several weeks under a regular light-dark cycle (14 hours of light and 10 hours of darkness per day).

The results of my studies of the body temperature and temperature selection rhythms in hamsters confirmed that the two rhythms are indeed 180°out of phase.[49] Having decided that the phenomenon was real, I wondered whether it occurred in other species as well. I found that it occurred in diurnal species (such as Chilean degus[50] and tree shrews[51]) as well as in nocturnal species (such as golden hamsters,[52] fat-tailed gerbils,[53] and flying squirrels[51]). The results from fat-tailed gerbils are shown in Figure 9. It is evident that, while body temperature peaks during the dark phase of the

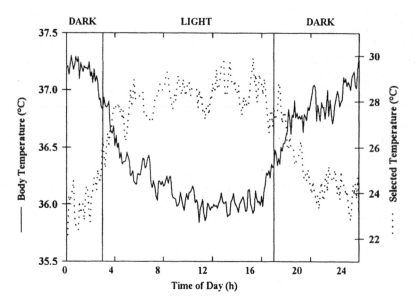

Figure 9 Average pattern of temperature selection and body temperature of fat-tailed gerbils allowed to select the temperature of their environment. (The curves show the average of various animals, each studied for many days.)

light-dark cycle, the rhythm of temperature selection peaks during the light phase, so that the two rhythms are exactly 180° out of phase.

Researchers in another laboratory later confirmed the finding of a preference for higher ambient temperature during the inactive phase in a nocturnal marsupial.[54] In my laboratory, I obtained evidence that the phenomenon occurs also in human subjects, although, of course, I could not keep the subjects in the experimental cage and had to employ a less stringent research protocol.[55]

Behavioral selection of ambient temperature is only one of many thermoregulatory responses available to mammals. Another response is *cold-induced thermogenesis* — that is, an increase in metabolic heat production in response to exposure to a cold environment. Would cold-induced thermogenesis be greater when the body temperature rhythm is at its low level? As indicated by my studies in hamsters[56] and fat-tailed gerbils,[53] the answer is *yes*. That is, the body temperature rhythm is opposed not only by the behavioral response of temperature selection but also — when such response is not available — by the autonomic response of cold-induced thermogenesis. Of course, if there really is an opposition, then the amplitude of the body temperature rhythm should be smaller when the animals are allowed to select their environmental temperature than when they are not. Although I failed to demonstrate this in the golden hamster, whose body temperature rhythm in a thermally stable environment has an amplitude of a little over 1°C,[49] I did succeed when I used flying squirrels and tree shrews, whose body temperature rhythms have a much greater amplitude.[51] The average amplitude for tree shrews maintained at a homogeneous temperature of 25°C was 4°C, while the average amplitude for tree shrews allowed to select their environmental temperature was 3°C.

To provide additional confirmation of the opposition between the thermoregulatory system and the circadian system, I examined the body temperature rhythm of hamsters after selective destruction of brain sites involved in the control of the two systems. Destruction of the circadian clock eliminated circadian rhythmicity but did not alter the main level of the body temperature rhythm[57-59] — which only confirms that the homeostatic and the circadian control of body temperature are exerted by separate brain sites. On the other hand, destruction of the main thermostat in the brain not only spared the body temperature rhythm but increased its amplitude.[59] I interpreted this finding as showing that, in intact animals, the brain thermostat causes a reduction in the amplitude of the oscillation imposed by the circadian clock. And I concluded from the data obtained in all the experiments that the circadian rhythm of body temperature in mammals is generated by the circadian system in opposition to the homeostatic stability defended by the thermoregulatory system.[60] Why such a functional conflict evolved is not clear. While a large decrease in body temperature during the inactive phase of the daily cycle can provide significant energetic savings (as is evident in animals exhibiting daily torpor),[61] the relatively small oscillation in body temperature observed in true homeotherms has no evident physiological function and, therefore, fails to provide an adaptive reason for the evolution of the circadian rhythm of body temperature. Based on recent findings about the effects of temperature on the cellular mechanisms of circadian rhythmicity, it has been suggested that the circadian rhythm of body temperature in mammals may serve the purpose of synchronizing subaltern autonomous clocks (such as the food-entrainable pacemaker) dispersed throughout the body.[62] While this suggestion requires further experimental support, it should be pointed out that the control of sleep also seems to involve separate circadian and homeostatic components that are often conflicting.[63]

Finally, we cannot complete our discussion of complex parameters of circadian rhythms without addressing the topic of *learning*. There is little doubt that the ability of the circadian pacemaker to be phase-shifted by environmental stimuli such as light is innate. However, can the circadian clock "learn" to phase shift in response to neutral stimuli previously associated with light? I tried unsuccessfully to make golden hamsters phase shift in response to sounds or odors previously associated with phase shifts caused by light pulses. I assume that other investigators also failed to demonstrate conditioning in the circadian system but did not publish their negative results. One group of investigators, however, reported that air puffs previously paired with light pulses can cause phase shifts in rats.[64] They also reported that conditioned fear can suppress the phase-shifting effect of light pulses.[65] If these results are replicated in other laboratories, the possibility of learning in the circadian system will have to be seriously contemplated.

bibliographic notes

1. Nelson, W., Tong, Y. L., Lee, J. K., and Halberg, F. (1979). Methods for cosinor rhythmometry. *Chronobiologia* 6: 305-323.
2. Aschoff, J. (1966). Circadian activity pattern with two peaks. *Ecology* 47: 657-662.
3. Refinetti, R. (1992). Non-parametric procedures for the determination of phase markers of circadian rhythms. *International Journal of Biomedical Computing* 30: 49-56.
4. Refinetti, R. (1992). Analysis of the circadian rhythm of body temperature. *Behavior Research Methods Instruments and Computers* 24: 28-36.
5. Refinetti, R. (1993). Comparison of six methods for the determination of the period of circadian rhythms. *Physiology and Behavior* 54: 869-875.
6. Basco, P. S., Rashotte, M. E., and Stephan, F. K. (1996). Photoperiod duration and energy balance in the pigeon. *Physiology and Behavior* 60: 151-159.
7. Refinetti, R. and Menaker, M. (1992). The circadian rhythm of body temperature of normal and tau-mutant golden hamsters. *Journal of Thermal Biology* 17: 129-133.
8. Refinetti, R. and Menaker, M. (1993). Independence of heart rate and circadian period in the golden hamster. *American Journal of Physiology* 264: R235-R238.
9. Pitrosky, B., Kirsch, R., Vivien-Roels, B., Georg-Bentz, I., Canguilhem, B., and Pévet, P. (1995). The photoperiodic response in Syrian hamster depends upon melatonin-driven circadian rhythm of sensitivity to melatonin. *Journal of Neuroendocrinology* 7: 889-895.
10. Moore-Ede, M. C. and Sulzman, F. M. (1981). Internal temporal order. In: Aschoff, J. (Ed.). *Handbook of Behavioral Neurobiology, Volume 4: Biological Rhythms*. New York: Plenum, pp. 215-241.
11. Refinetti, R. (1996). Comparison of the body temperature rhythms of diurnal and nocturnal rodents. *Journal of Experimental Zoology* 275: 67-70.
12. Altman, P. L. and Dittmer, D. S. (1968). *Biological Handbook: Metabolism*. Bethesda, MD: FASEB.
13. Dawson, W. R. and Hudson, J. W. (1970). Birds. In: Whittow, G. C. (Ed.). *Comparative Physiology of Thermoregulation, Volume 1*. New York: Academic, pp. 223-310.
14. Tyndale-Biscoe, H. (1973). *Life of Marsupials*. London: Edward Arnold.
15. Aschoff, J., Gerecke, U., and Wever, R. (1967). Phasenbeziehungen zwischen den circadianen Perioden der Aktivität und der Kerntemperatur beim Menschen. *Pflügers Archiv* 295: 173-183.
16. Aschoff, J. and Wever, R. (1976). Human circadian rhythms: A multioscillatory system. *Federation Proceedings* 35: 2326-2332.
17. Kriebel, J. (1974). Changes in internal phase relationships during isolation. In: Scheving, L. E., Halberg, F., and Pauly, J. E. (Eds.). *Chronobiology*. Tokyo: Igaku Shoin, pp. 451-459.
18. Weitzman, E. D., Moline, M. L., Czeisler, C. A., and Zimmerman, J. C. (1982). Chronobiology of aging: Temperature, sleep-wake rhythms and entrainment. *Neurobiology of Aging* 3: 299-309.
19. Refinetti, R. (1997). Phase relationship of the body temperature and locomotor activity rhythms in free-running and entrained rats. *Biological Rhythm Research* 28 (Suppl.): 19-24.

20. Refinetti, R. (1999). Relationship between the daily rhythms of locomotor activity and body temperature in eight mammalian species. *American Journal of Physiology* 277: R00-R00.
21. Maurel, E. (1884). Expériences sur les variations nycthémérales de la température normale. *Comptes Rendus des Séances de la Société de Biologie de Strasbourg* 37: 588.
22. Grande, F., Monagle, J. E., Buskirk, E. R., and Taylor, H. L. (1959). Body temperature responses to exercise in man on restricted food and water intake. *Journal of Applied Physiology* 14: 194-198.
23. Saltin, G., Gagge, A. P., and Stolwijk, J. A. J. (1970). Body temperatures and sweating during thermal transients caused by exercise. *Journal of Applied Physiology* 28: 318-327.
24. Bleichert, A., Behling, K., Scarperi, M., and Scarperi, S. (1973). Thermoregulatory behavior of man during rest and exercise. *Pflügers Archiv* 338: 303-312.
25. Fujishima, K. (1986). Thermoregulatory responses during exercise and a hot water immersion and the affective responses to peripheral thermal stimuli. *International Journal of Biometereology* 30: 1-19.
26. Thompson, G. E. and Stevenson, J. A. F. (1965). The temperature response of the male rat to treadmill exercise, and the effect of anterior hypothalamic lesions. *Canadian Journal of Physiology and Pharmacology* 43: 279-287.
27. Campbell, B. A. and Lynch, G. S. (1968). Influence of hunger and thirst on the relationship between spontaneous activity and body temperature. *Journal of Comparative and Physiological Psychology* 65: 492-498.
28. Kluger, M. J., Gonzalez, R. R., and Stolwijk, J. A. J. (1973). Temperature regulation in the exercising rabbit. *American Journal of Physiology* 224: 130-135.
29. Paladino, F. V. and King, J. R. (1984). Thermoregulation and oxygen consumption during terrestrial locomotion by white-crowned sparrows Zonotrichia leucophrys gambelii. *Physiological Zoology* 57: 226-236.
30. Murray, E. J., Williams, H. L., and Lubin, A. (1958). Body temperature and psychological ratings during sleep deprivation. *Journal of Experimental Psychology* 56: 271-273.
31. Smith, R. E. (1969). Circadian variations in human thermoregulatory responses. *Journal of Applied Physiology* 26: 554-560.
32. Marotte, H. and Timbal, J. (1981). Circadian rhythm of temperature in man: Comparative study with two experiment protocols. *Chronobiologia* 8: 87-100.
33. Gander, P. H., Connell, L. J., and Graeber, R. C. (1986). Masking of the circadian rhythms of heart rate and core temperature by rest-activity cycle in man. *Journal of Biological Rhythms* 1: 119-135.
34. Bolles, R. C., Duncan, P. M., Grossen, N. E., and Matter, C. F. (1968). Relationship between activity level and body temperature in the rat. *Psychological Reports* 23: 991-994.
35. Honma, K. and Hiroshige, T. (1978). Simultaneous determination of circadian rhythms of locomotor activity and body temperature in the rat. *Japanese Journal of Physiology* 28: 159-169.
36. Refinetti, R. (1994). The contribution of locomotor activity to the generation of the daily rhythm of body temperature in golden hamsters. *Physiology and Behavior* 56: 829-831.
37. Crawshaw, L. I. (1974). Temperature selection and activity in the crayfish, *Orconectes immunis*. *Journal of Comparative Physiology* 95: 315-322.

38. Reynolds, W. W., Casterlin, M. E., Matthey, J. K., Millington, S. T., and Ostrowski, A. C. (1978). Diel patterns of preferred temperature and locomotor activity in the goldfish *Carassius auratus. Comparative Biochemistry and Physiology* 59A: 225-227.

39. Kavaliers, M. and Ralph, C. L. (1980). Pineal involvement in the control of behavioral thermoregulation of the white sucker, *Catostomus commersoni. Journal of Experimental Zoology* 212: 301-303.

40. Regal, P. J. (1967). Voluntary hypothermia in reptiles. *Science* 155: 1551-1553.

41. Refinetti, R. and Susalka, S. J. (1997). Circadian rhythm of temperature selection in a nocturnal lizard. *Physiology and Behavior* 62: 331-336.

42. Cowgell, J. and Underwood, H. (1979). Behavioral thermoregulation in lizards: A circadian rhythm. *Journal of Experimental Zoology* 210: 189-194.

43. Innocenti, A., Minutini, L., and Foà, A. (1993). The pineal and circadian rhythms of temperature selection and locomotion in lizards. *Physiology and Behavior* 53: 911-915.

44. Tosini, G. and Menaker, M. (1995). Circadian rhythm of body temperature in an ectotherm (*Iguana iguana*). *Journal of Biological Rhythms* 10: 248-255.

45. Briese, E. (1985). Rats prefer ambient temperature out of phase with their body temperature circadian rhythm. *Brain Research* 345: 389-393.

46. Briese, E. (1986). Circadian body temperature rhythm and behavior of rats in thermoclines. *Physiology and Behavior* 37: 839-847.

47. Gordon, C. J. (1993). Twenty-four hour rhythms of selected ambient temperature in rat and hamster. *Physiology and Behavior* 53: 257-263.

48. Gordon, C. J. (1994). 24-hour control of body temperature in rats. I. Integration of behavioral and autonomic effectors. *American Journal of Physiology* 267: R71-R77.

49. Refinetti, R. (1995). Rhythms of temperature selection and body temperature are out of phase in the golden hamster. *Behavioral Neuroscience* 109: 523-527.

50. Refinetti, R. (1996). Rhythms of body temperature and temperature selection are out of phase in a diurnal rodent, *Octodon degus. Physiology and Behavior* 60: 959-961.

51. Refinetti, R. (1998). Body temperature and behavior of tree shrews and flying squirrels in a thermal gradient. *Physiology and Behavior* 63: 517-520.

52. Refinetti, R. (1995). Body temperature and behavior of golden hamsters (*Mesocricetus auratus*) and ground squirrels (*Spermophilus tridecemlineatus*) in a thermal gradient. *Journal of Comparative Physiology A* 177: 701-705.

53. Refinetti, R. (1998). Homeostatic and circadian control of body temperature in the fat-tailed gerbil. *Comparative Biochemistry and Physiology* 119A: 295-300.

54. Song, X., Körtner, G., and Geiser, F. (1998). Temperature selection and use of torpor by the marsupial *Sminthopsis macroura. Physiology and Behavior* 64: 675-682.

55. Shoemaker, J. A. and Refinetti, R. (1996). Day-night difference in the preferred ambient temperature of human subjects. *Physiology and Behavior* 59: 1001-1003.

56. Watts, R. H., Jr. and Refinetti, R. (1996). Circadian modulation of cold-induced thermogenesis in the golden hamster. *Biological Rhythm Research* 27: 87-94.

57. Refinetti, R., Kaufman, C. M., and Menaker, M. (1994). Complete suprachiasmatic lesions eliminate circadian rhythmicity of body temperature and locomotor activity in golden hamsters. *Journal of Comparative Physiology A* 175: 223-232.

58. Refinetti, R. (1995). Effects of suprachiasmatic lesions on temperature regulation in the golden hamster. *Brain Research Bulletin* 36: 81-84.

59. Osborne, A. R. and Refinetti, R. (1995). Effects of hypothalamic lesions on the body temperature rhythm of the golden hamster. *NeuroReport* 6: 2187-2192.

60. Refinetti, R. (1997). Homeostasis and circadian rhythmicity in the control of body temperature. *Annals of the New York Academy of Sciences* 813: 63-70.

61. Lyman, C. P., Willis, J. S., Malan, A., and Wang, L. C. H. (1982). *Hibernation and Torpor in Mammals and Birds.* New York: Academic.

62. Liu, Y., Merrow, M., Loros, J. J., and Dunlap, J. C. (1998). How temperature changes reset a circadian oscillator. *Science* 281: 825-829.

63. Heller, H. C., Edgar, D. M., Grahn, D. A., and Glotzbach, S. F. (1996). Sleep, thermoregulation, and circadian rhythms. In: Fregly, M. J. and Blatteis, C. M. (Eds.). *Handbook of Physiology, Section 4: Environmental Physiology.* New York: Oxford University Press, v. 2, pp. 1361-1374.

64. Amir, S. and Stewart, J. (1996). Resetting of the circadian clock by a conditioned stimulus. *Nature* 379: 542-545.

65. Amir, S. and Stewart, J. (1998). Conditioned fear suppresses light-induced resetting of the circadian clock. *Neuroscience* 86: 345-351.

part III

---

circadian pacemaker: neuroscience and molecular biology

chapter eight

physical location of the clock

Following the discussion of behavioral and physiological aspects of circadian rhythms in the preceding chapters, the physical substrate of the circadian clock (the suprachiasmatic nucleus of the hypothalamus) is addressed here.

Once you know that there is a clock somewhere in the body, how do you go about locating this clock? There is no easy answer for this question. As the brain controls most bodily functions, one would be disposed to look at the brain first. But where in the brain? The brain contains about 100 billion nerve cells, which renders the trial-and-error approach useless. Maybe luck? As a matter of fact, the discovery of a "pleasure center" in the brain by James Olds some 40 years ago[1] was prompted by sheer luck. When Olds was a graduate student, he was assigned the job of implanting electrodes (very thin pieces of conducting wire) in the brains of laboratory rats for his professor's research. Of course, the electrodes were implanted under anesthesia and were so small that they did not bother the animals afterwards. The goal was to pass minute electrical currents to simulate the natural electrical activity of nerve cells. As a beginner, Olds was so inaccurate in the placement of the electrodes that he ended up placing them almost on the opposite side of the brain from that intended by his professor. It ruined the professor's study. However, it resulted in a pleasant surprise: It turned out that direct electrical stimulation of that part of the brain (the lateral hypothalamus) resulted in the closest thing to "ultimate pleasure" that a rat could experience. Rats would spend hours performing a dull task (depressing a lever on the cage wall) if this behavior provided them with stimulation of the lateral hypothalamus. They would cross a highly charged electrical grid in order to reach the lever and be able to self-stimulate their brains. They would forgo food or a female in heat in order to self-stimulate. Thus, Olds seemed to have found the "pleasure center" in the brain. By pure luck, he made one of the first discoveries about neural mechanisms involved in motivation. Much research immediately followed his discovery and is still ongoing today.

The discovery of the anatomical site of the circadian clock in mammals also depended on unplanned actions and the ability to integrate fragmented pieces of evidence. Although much less was known about circadian rhythms in the early 1970s than is known now, researchers were confident that a circadian clock would be found somewhere in the body. As we saw in Chapters 4 and 5, the study of circadian rhythms had indicated the necessity of the existence of a circadian pacemaker. Researchers also knew that the pacemaker was sensitive to light, although they had no idea whether the pacemaking property of the clock and its sensitivity to light had a single physical substrate or two separate substrates.

Fredric Stephan, now a professor at Florida State University, was then a graduate student in Dr. Irving Zucker's laboratory at the University of California, Berkeley. Stephan's original interest was in reproductive physiology and behavior. Zucker's main interest, however, was in circadian rhythms, and one need not guess what kind of research Stephan ended up doing. His interest in reproductive physiology came out handy, though. It happened that reproductive physiologists at the time had already found out that the estrous cycle in rats (the equivalent of the menstrual cycle in humans) is dependent on a small part of the brain, namely, the rostral (frontal) part of the hypothalamus. When a dependence of the estrous cycle on circadian rhythms was established,[2] Stephan put A and B together and wondered whether the circadian pacemaker might be located in the rostral hypothalamus. Yet, the rostral hypothalamus is not that small. Where in the rostral hypothalamus could the circadian pacemaker be?

The final hint came from findings at Dr. Robert Y. Moore's laboratory at the University of Pittsburgh. Moore had been studying the nervous pathways that take information about light from the eyes to the brain. One pathway was already well known: the one related to vision. What was not clear was whether there would be a separate pathway responsible for the entrainment of circadian rhythms. What Moore found out—and Stephan heard about it even before publication — was that there was indeed a second path from the retina of the eyes to the rostral hypothalamus, which was accordingly named the retino-hypothalamic tract.[3] Where did the retino-hypothalamic tract end? At a small (a few thousand cells) bilateral nucleus at the base of the brain called the *suprachiasmatic nucleus*. As shown in Figure 10, the suprachiasmatic nucleus (or nuclei, since there are two of them) is so called because it sits right above the optic chiasm (the crossing of the optic nerves). So, that was it — the suprachiasmatic nucleus (SCN) had to be the physical substrate of the circadian pacemaker.

Now, once you have a suspicion, how do you proceed? Well, if a brain structure houses the biological clock, destruction of this structure must eliminate biological timing. Right, but how can you destroy a minute part of the brain of a living animal without damaging everything around it? This is a question that researchers had faced in the late 1800s when they started to systematically study the effects of the destruction of different brain areas on a variety of physiological processes (such as hunger, thirst, temperature

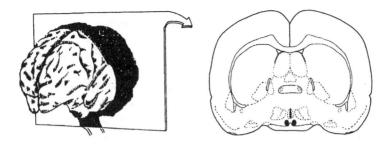

Figure 10 Diagram of the brain showing the location of the suprachiasmatic nuclei. The suprachiasmatic nuclei are the two olive shaped structures at the base of the brain, as seen in a coronal section of the rat brain (*right*). A coronal section is a transverse cut that divides the brain into a front and a back portion (*left*).

regulation, etc.). By the 1950s, procedures for selectively lesioning small areas of the brain were well established. Indeed, when my former graduate advisor — Dr. Harry Carlisle — finished graduate school in the 1960s, he used the lesioning technique to study the role of the rostral hypothalamus in temperature regulation in the rat.[4] As a matter of fact, when I was in graduate school in the mid-1980s, I used the same technique to study the role of the lateral hypothalamus in temperature regulation.[5]

Although this type of brain surgery, like any type of brain surgery, is certainly not something simple, it requires only three basic elements — a stereotaxic atlas, a stereotaxic instrument, and a lesioning electrode. A stereotaxic atlas is simply a map of the brain that shows where each structure is in relation to the skull. A stereotaxic instrument is a precision instrument that allows the investigator to reach structures in the brain with millimetric resolution through a tiny hole drilled in the skull. A lesioning electrode is an electrode (thin wire) used to lesion the brain tissue, usually by passage of moderate electric current. Thus, the atlas tells you where the particular brain structure is; the stereotaxic instrument allows you to reach the structure without damaging the surrounding tissue; and the electrode provides the means to lesion the structure.

So, armed with the suspicion that the circadian pacemaker might be located in the SCN, and the knowledge of how to destroy the SCN without damaging other parts of the brain, Stephan went to work on his project. After many of the usual setbacks and partial accomplishments, he eventually found that complete lesions of the SCN (but not partial ones) eliminated the circadian rhythms of behavioral activity and drinking in rats. Hooray! He had found the physical substrate of the biological clock! A great achievement that, he thought, deserved to be publicized in the top science journal, *Science*. As it turned out, the editors of *Science* were not that impressed with his findings and declined to publish his article. The article was published in the *Proceedings of the National Academy of Sciences* (a very reputable publication) in 1972.[6] Ironically, Stephan has since been studying the "food-entrainable pacemaker" (see Chapter 6), whose neural substrate he has never managed to identify.

Not surprisingly, Moore's team did not fail to infer the role of the SCN on their own; they published results similar to Stephan's in the same year of 1972.[7] Elimination of the circadian rhythms of activity, drinking, and various other rhythms (such as body temperature, blood hormone levels, sleep, and oscillations in heart rate and blood pressure) by complete lesions of the SCN (but not by lesions of other brain areas) has been repeatedly demonstrated since then in a variety of mammalian species.[8-16] I myself have abolished the rhythms of activity and body temperature by SCN lesions in hamsters.[17-19] In birds and lizards, elimination of circadian rhythmicity has also been observed after destruction of brain structures analogous to the mammalian SCN.[20-22] But wait a minute! Anyone who has heard the anecdote about the legless flea knows that this evidence is not enough. The story goes somewhat like this:

> A flea trainer conducted an experiment with one of his jumping fleas. He approached the flea and said "Jump, flea"— and the flea jumped. So, he wrote down, "A healthy, six-legged flea will jump on command." Then, he removed one of the flea's legs and said "Jump, flea"— and the flea obliged. The amateur experimenter proceeded meticulously, removing a leg at a time and observing the results. Finally, when the flea had no legs left, it did not jump on command. The flea trainer wrote down: "A flea without legs is deaf."

Obviously, removal of the legs had nothing to do with the flea's ability to hear the commands. The flea could hear perfectly well. It just did not have the legs to jump. Likewise, the SCN might be simply part of the nervous pathway that connects the clock to the effector organs. If the SCN really contains a clock, it should display rhythmicity in isolation. Can we remove the SCN from the brain and study it in a dish? Actually, we can. Electrophysiologists had been studying hypothalamic tissue *in vitro* (Latin for "in a dish") for years when circadian biologists decided to record the electrical activity of SCN cells from the rat brain. Brain slices cannot be studied for several weeks, as whole animals can, but clear circadian rhythms in the combined electrical activity of SCN cells were recorded continuously for several days.[23-25] Even though rats are active at night, the SCN cells were found to be more active during what would have been the daytime had the animals not been killed.

So, destruction of the SCN in whole animals abolishes circadian rhythmicity, and SCN slices *in vitro* exhibit rhythmicity. Is this proof enough that the SCN is the neural substrate of the circadian pacemaker? Not if you are a compulsive skeptic, as most scientists are. The brain-slice preparation is ostensibly unnatural; would the SCN display rhythmicity *in vivo* (that is, in the living animal)? This has been investigated in two ways. Some researchers replicated the *in vitro* studies *in vivo*. By using chronically implanted recording electrodes, they recorded the electrical activity of SCN tissue in freely moving guinea pigs and rats and found robust circadian rhythmicity.[26–27] The SCN cells were found to be more active during the daytime, whereas the rest of the brain (as well as the whole animal) was more active during the nighttime. Other researchers measured SCN activity *in vivo* using the 2-deoxy-glucose methodology. The 2-deoxy-glucose (2-DG, for short) methodology had been developed by Dr. Louis Sokoloff in the 1970s.[28] The methodology involves the use of radioactively labeled 2-DG, which is an analogue of glucose. Brain cells utilize glucose as their main substrate for energy metabolism, so that cells that are more active utilize more glucose. While 2-DG is utilized similarly to glucose, it is not fully metabolized and gets trapped in the tissues. Therefore, more active cells accumulate more 2-DG. By radiographic techniques, one can determine what parts of the brain were more active following a 2-DG injection, as they accumulate more radioactive molecules. The use of this methodology allowed investigators to determine that the SCN of freely moving rats exhibits circadian rhythmicity, being more active during the daytime (while the rest of the brain and the whole animal are more active at nighttime).[29]

As if all of this were not enough to establish beyond doubt that the SCN is the site of the circadian pacemaker, researchers sought even further evidence. If a brain structure houses the biological clock, not only should destruction of this structure eliminate rhythmicity, but replacement of the structure should restore it. Now, mature brain tissue cannot be successfully transplanted, but fetal brain tissue can be grafted onto an adult brain. Thus, when adult hamsters rendered arrhythmic by SCN lesions received implants of hypothalamic tissue from fetal hamsters or immortalized SCN cell lines, circadian rhythmicity was restored.[30–33] Of course, since the circadian periods of the donor and the host are very similar, one cannot be sure that the implant actually replaced the clock (rather than provided an essential trophic factor required by some remaining portion of the host's clock). To eliminate the doubt, other researchers transplanted fetal tissue from one species to SCN-lesioned adults of a different species: from rats to chipmunks[34] and from rats and mice to hamsters.[35] Again, rhythmicity was restored, and the period of the rhythm matched, at least closely, that of the donor — not that of the host.

Even more dramatic were the results of transplants utilizing the *tau* mutant hamster. As described in Chapter 5, these hamsters are genetically set to circadian periods of 20, 22, and 24 hours. What would happen if an adult hamster with a period of 24 hours were to be rendered arrhythmic by SCN lesion and were implanted with the fetal SCN of a hamster with a period of 20 hours? Yes, you guessed it right — experimental results showed that the adult hamster that had previously had a 24-hour period started to exhibit, after a few days of recovery from surgery, 20-hour rhythmicity.[36] The converse was also true: Lesioned hamsters with an original period of 20 hours started to exhibit 24-hour rhythmicity after receiving a hypothalamic implant from fetuses of the 24-hour-period genotype. Furthermore, when animals received implants after having experienced only partial lesion of the SCN, they showed 20-hour and 24-hour rhythmicity at the same time.[37]

Based on all the experimental evidence described above, it seems justified to conclude that the suprachiasmatic nucleus is the physical substrate of the circadian pacemaker. As we look in more detail into the neurophysiology and pharmacology of the clock in the next chapter, however, we will have to place some restrictions in our conclusion. For instance, a few other parts of the brain — particularly in birds and other non-mammalian vertebrates — seem to operate as secondary clocks. Also, the SCN cannot possibly be the anatomical site of the circadian clock in invertebrates, as these do not possess an SCN. Interestingly, however, we will see in Chapter 10 later on that the molecular mechanism of the circadian clock seems to be very similar in all life forms — from plants and fungi to mammals.

bibliographic notes

1. Olds, J. (1958). Self-stimulation of the brain. *Science* 127: 315-324.
2. Alleva, J. J., Waleski, M. V., and Alleva, F. R. (1971). A biological clock controlling the estrous cycle of the hamster. *Endocrinology* 88: 1368-1379.
3. Moore, R. Y. and Lenn, N. J. (1972). A retinohypothalamic projection in the rat. *Journal of Comparative Neurology* 146: 1-14.
4. Carlisle, H. J. (1969). Effect of preoptic and anterior hypothalamic lesions on behavioral thermoregulation in the cold. *Journal of Comparative and Physiological Psychology* 69: 391-402.
5. Refinetti, R. and Carlisle, H. J. (1986). Effects of lateral hypothalamic lesions on thermoregulation in the rat. *Physiology and Behavior* 38: 219-228.
6. Stephan, F. K. and Zucker, I. (1972). Circadian rhythm in drinking behavior and locomotor activity of rats are eliminated by hypothalamic lesions. *Proceedings of the National Academy of Sciences* 69: 1583-1586.
7. Moore, R. Y. and Eichler, V. B. (1972). Loss of a circadian adrenal corticosterone rhythm following suprachiasmatic lesions in the rat. *Brain Research* 42: 201-206.
8. Stephan, F. K. and Nunez, A. A. (1977). Elimination of circadian rhythms in drinking, activity, sleep, and temperature by isolation of the suprachiasmatic nuclei. *Behavioral Biology* 20: 1-16.
9. Warren, W. S., Champney, T. H., and Cassone, V. M. (1994). The suprachiasmatic nucleus controls the circadian rhythm of heart rate via the sympathetic nervous system. *Physiology and Behavior* 55: 1091-1099.
10. Sano, H., Hayashi, H., Makino, M., Takezawa, H., Hirai, M., Saito, H., and Ebihara, S. (1995). Effects of suprachiasmatic lesions on circadian rhythms of blood pressure, heart rate, and locomotor activity in the rat. *Japanese Circulation Journal* 59: 565-573.
11. Ruby, N. F., Ibuka, N., Barnes, B. M., and Zucker, I. (1989). Suprachiasmatic nuclei influence torpor and circadian temperature rhythms in hamsters. *American Journal of Physiology* 257: R210-R215.
12. Stetson, M. H. and Watson-Whitmyre, M. (1976). Nucleus suprachiasmaticus: The biological clock in the hamster? *Science* 191: 197-199.

13. Aalto, J. and Kiianmaa, K. (1984). Circadian rhythms of water and alcohol intake: Effect of REM-sleep deprivation and lesion of the suprachiasmatic nucleus. *Alcohol* 1: 403-407.
14. Zucker, I., Boshes, M., and Dark, J. (1983). Suprachiasmatic nuclei influence circannual and circadian rhythms of ground squirrels. *American Journal of Physiology* 244: R472-R480.
15. Sato, T. and Kawamura, H. (1984). Effects of bilateral suprachiasmatic nucleus lesions on the circadian rhythms in a diurnal rodent, the Siberian chipmunk. *Journal of Comparative Physiology A* 155: 745-752.
16. DeCoursey, P. J., Krulas, J. R., Mele, G., and Holley, D. C. (1997). Circadian performance of suprachiasmatic nuclei (SCN)-lesioned antelope ground squirrels in a desert enclosure. *Physiology and Behavior* 62: 1099-1108.
17. Refinetti, R., Kaufman, C. M., and Menaker, M. (1994). Complete suprachiasmatic lesions eliminate circadian rhythmicity of body temperature and locomotor activity in golden hamsters. *Journal of Comparative Physiology A* 175: 223-232.
18. Refinetti, R. (1995). Effects of suprachiasmatic lesions on temperature regulation in the golden hamster. *Brain Research Bulletin* 36: 81-84.
19. Osborne, A. R. and Refinetti, R. (1995). Effects of hypothalamic lesions on the body temperature rhythm of the golden hamster. *NeuroReport* 6: 2187-2192.
20. Takahashi, J. S. and Menaker, M. (1982). Role of the suprachiasmatic nuclei in the circadian system of the house sparrow, *Passer domesticus*. *Journal of Neuroscience* 2: 815-828.
21. Janik, D. S., Pickard, G. E., and Menaker, M. (1990). Circadian locomotor rhythms in the desert iguana. II. Effects of electrolytic lesions to the hypothalamus. *Journal of Comparative Physiology A* 166: 811-816.
22. Minutini, L., Innocenti, A., Bertolucci, C., and Foà, A. (1995). Circadian organization in the ruin lizard *Podarcis sicula*: The role of the suprachiasmatic nuclei of the hypothalamus. *Journal of Comparative Physiology A* 176: 281-288.
23. Satinoff, E., Li, H., Tcheng, T. K., Liu, C., McArthur, A. J., Medanic, M., and Gillette, M. U. (1993). Do the suprachiasmatic nuclei oscillate in old rats as they do in young ones? *American Journal of Physiology* 265: R1216-R1222.
24. Prosser, R. A., Edgar, D. M., Heller, H. G., and Miller, J. D. (1994). A possible glial role in the mammalian circadian clock. *Brain Research* 643: 296-301.
25. Moriya, T., Fukushima, T., Shimazoe, T., Shibata, S., and Watanabe, S. (1996). Chronic administration of methamphetamine does not affect the suprachiasmatic nucleus-operated circadian pacemaker in rats. *Neuroscience Letters* 208: 129-132.
26. Kurumiya, S. and Kawamura, H. (1988). Circadian oscillation of the multiple unit activity in the guinea pig suprachiasmatic nucleus. *Journal of Comparative Physiology A* 162: 301-308.
27. Meijer, J. H., Watanabe, K., Schaap, J., Albus, H., and Détári, L. (1998). Light responsiveness of the suprachiasmatic nucleus: Long-term multiunit and single-unit recordings in freely moving rats. *Journal of Neuroscience* 18: 9078-9087.
28. Sokoloff, L. (1981). Localization of functional activity in the central nervous system by measurement of glucose utilization with radioactive deoxyglucose. *Journal of Cerebral Blood Flow and Metabolism* 1: 7-36.
29. Room, P. and Tielemans, A. J. P. C. (1989). Circadian variations in local cerebral glucose utilization in freely moving rats. *Brain Research* 505: 321-325.

30. DeCoursey, P. J. and Buggy, J. (1988). Restoration of circadian locomotor activity in arrhythmic hamsters by fetal SCN transplants. *Comparative Endocrinology* 4: 49-54.
31. Romero, M. T., Lehman, M. N., and Silver, R. (1993). Age of donor influences ability of suprachiasmatic nucleus grafts to restore circadian rhythmicity. *Developmental Brain Research* 71: 45-52.
32. Le Sauter, J. and Silver, R. (1994). Suprachiasmatic nucleus lesions abolish and fetal grafts restore circadian gnawing rhythms in hamsters. *Restorative Neurology and Neuroscience* 6: 135-143.
33. Earnest, D. J., Liang, F. Q., Ratcliff, M., and Cassone, V. M. (1999). Immortal time: Circadian clock properties of rat suprachiasmatic cell lines. *Science* 283: 693-695.
34. Saitoh, Y., Matsui, Y., Nihonmatsu, I., and Kawamura, H. (1991). Cross-species transplantation of the suprachiasmatic nuclei from rats to Siberian chipmunks with suprachiasmatic lesions. *Neuroscience Letters* 123: 77-81.
35. Sollars, P. J., Kimble, D. P., and Pickard, G. E. (1995). Restoration of circadian behavior by anterior hypothalamic heterografts. *Journal of Neuroscience* 15: 2109-2122.
36. Ralph, M. R., Foster, R. G., Davis, F. C., and Menaker, M. (1990). Transplanted suprachiasmatic nucleus determines circadian period. *Science* 247: 975-978.
37. Vogelbaum, M. A. and Menaker, M. (1992). Temporal chimeras produced by hypothalamic transplants. *Journal of Neuroscience* 12: 3619-3627.

chapter nine

neurophysiology and pharmacology of the clock

> This chapter examines the neural organization of the suprachiasmatic nucleus and its afferent and efferent connections, including the neurotransmitters used for synaptic communication.

In the preceding chapter, we saw that the circadian pacemaker in mammals is located in the suprachiasmatic nuclei (SCN) of the hypothalamus. Destruction of the SCN eliminates circadian rhythms, and transplanted grafts restore them. Also, SCN tissue displays circadian rhythmicity *in vitro* and *in vivo*. Having examined this evidence, it is now time to look closer at the organization of the SCN and the pathways that take information to it and from it to all other organs in the body. I will concentrate on mammalian biology and will point out some major differences from the biology of other vertebrates. The circadian system of invertebrates has similarities with that of vertebrates but is also quite different. Circadian rhythms in invertebrates, as in vertebrates, are controlled by the nervous and neuroendocrine systems; however, the invertebrate nervous system is more malleable than that of vertebrates, and the invertebrate circadian system seems to involve multiple, powerful circadian pacemakers located in the central nervous system as well as in the eyes and other structures.[1]

pacemaking cells

In principle, the SCN could generate circadian rhythmicity either by relying on rhythmic cells or by structuring the activity of cells that do not exhibit circadian rhythmicity themselves (a process of "emergent" rhythmicity). Although it has been indicated that circadian rhythmicity could be derived by the coordination of the activity of multiple ultradian oscillators,[2, 3] recent studies have clearly shown that individual SCN cells display circadian rhythmicity.

The fact that animals with extensive SCN lesions, which spare only a few cells, display circadian rhythmicity is strong evidence that rhythmicity must be a cellular process.[4] Of course, real proof comes only from direct studies of the activity of single SCN cells. *In vitro* studies of the electrical activity of single cells in SCN slices or cell cultures from rodent brain have clearly established that individual cells exhibit circadian rhythmicity, that different cells have different circadian periods, and that the overall period of the circadian oscillation generated by the SCN is the average of the period of the various cells.[5-9] The range of periods of individual cells is much wider than that of the whole SCN. For instance, in the rat, whose circadian period is approximately 24.2 hours, the standard deviation of the mean of individual cells is 1.4 hours whereas the standard deviation of the mean of the locomotor activity rhythm of the animals is 0.2 hour.[8] However, this difference is easily explained by basic statistical properties of the arithmetic mean. Since the standard deviation of the distribution of means equals the standard deviation of the distribution of the population divided by the square root of the group size ($\sigma_M = \sigma_x/n^{0.5}$), it can be predicted that the mean period of a group of as few as 50 cells will have a standard deviation of 0.2 hour. As a matter of fact, based on the standard deviation of the individual cells (1.4 hours), one would expect the group of 10,000 SCN cells to have a mean period with a standard deviation slightly larger than 0.01 hour. Obviously, some source of biological noise must be responsible for the increase in the standard deviation from 0.01 to 0.2 hour.

The cellular processes responsible for the generation of circadian rhythmicity are still unknown (molecular mechanisms will be discussed in Chapter 10). However, much has been learned recently about the neurotransmitters used for communication between nerve cells in the SCN. Three neurotransmitters are particularly abundant in the SCN — GABA (gamma-aminobutyric acid, a ubiquitous inhibitory aminoacid transmitter found throughout the central nervous system), AVP (arginine vasopressin, a neurotransmitter that, as indicated by its name, was first identified by its role in the control of dilation and constriction of blood vessels), and VIP (vasoactive intestinal peptide, a peptide first identified by its action on the intestinal wall).

Anatomical studies using immunocytochemical methods (that is, methods that utilize especially prepared antibodies that contain dyes or radioactive substances and that attach to the neurotransmitter in an antigen-antibody reaction) have demonstrated the presence of GABA neurons throughout the SCN.[10-12] GABA administration in hypothalamic slices *in vitro* was shown to inhibit the activity of approximately 90% of SCN cells.[13] While the sensitivity of SCN cells to GABA administration was originally reported to be constant across the circadian cycle,[14] administration of baclofen (a GABA agonist) caused phase-dependent phase shifts in the rhythm of electrical activity of SCN cells *in vitro*.[15] GABA administration to rat brain slices was found to be inhibitory at night but excitatory during the day.[16] Also, drugs that affect the activity of GABA receptors (in particular, benzodiazepines) have been shown to cause phase shifts of behavioral rhythms and changes in free-running period when administered systemically.[17-20] In addition, administration of a GABA precursor (glutamate) to individual SCN cells *in vitro* caused consistent alterations in membrane potential, which supports the idea that SCN neurons receive GABA-mediated synaptic input from other SCN neurons.[21]

AVP is also abundant in the SCN, but it is restricted to the dorsomedial section of the nucleus.[10,11,22,23] Functional studies *in vitro* clearly showed that AVP is released from synaptic terminals at a higher rate during the day than during the night in rats maintained under a light-dark cycle and that the release remains rhythmic in animals maintained in constant darkness.[23-25] (As mentioned in Chapter 8, the electrical activity of the SCN is higher during the day than during the night in both diurnal and nocturnal animals, even though the activity of the rest of the brain, and of the whole animal, is higher during the night in nocturnal animals.) Interestingly, a study in voles indicated that animals with a smaller number of AVP immunoreactive cells have more robust rhythms of running-wheel activity than those with a greater number of such cells.[26] In contrast to the AVP-release data, electrophysiological and immunoreactive studies showed that the sensitivity of SCN cells to AVP in nocturnal rodents is greater during the night than during the day.[27-29] Presumably, the phase difference between AVP release and sensitivity results in a flat overall activation of AVP synapses across the circadian cycle.

In contrast to AVP, VIP is present mostly in the ventrolateral section of the SCN.[10,22,23,30,31] (As described below, most of the afferent pathways to the SCN end in the ventrolateral section.) VIP receptors, however, have been found to be present throughout the SCN and to not display circadian oscillation in sensitivity[32] or to display greater sensitivity during the day.[29] VIP administration *in vivo* has been found to cause phase shifts of circadian rhythms similar to those caused by light pulses.[33]

Evidence in support of other neurotransmitters utilized in the communication among SCN cells has been produced, although additional studies are necessary to further document the findings. These neurotransmitters include norepinephrine,[34, 35] somatostatin,[23,25] and gastrin releasing peptide.[22,29,33]

afferent pathways

Since the circadian pacemaker is responsive to light, the SCN must receive information about photic stimuli. In birds, fish, and reptiles, the eyes and the pineal gland (and, sometimes, additional structures) contain photoreceptors that can transduce electromagnetic energy into nerve impulses (or, more accurately, into hyperpolarization of the receptors, which later leads to the generation of nerve impulses).[36] In mammals, the eyes are the only sensory organs capable of transducing photic stimuli. To be exact, it has been recently reported that circadian rhythms in humans can be phase shifted by photic stimulation of the small area behind the knee joint.[37] If this finding is replicated, it will force us to assume that some organ in the body is sensitive to changes in blood substances elicited by exposure to light. For the time being, it can be said that all photic stimulation reaching the SCN originates in the eyes. This is the case because circadian rhythms of animals whose eyes have been removed do not respond to photic stimulation,[38-40] while circadian rhythms of animals that have eyes but are visually blind do respond to photic stimulation.[41-43] In rats and hamsters, the spectral sensitivity of the circadian system (as indicated by phase shifts or entrainment produced by light of different colors) matches the spectral sensitivity of the visual pigment rodopsin,[44,45] which suggests that retinal rods may be the photic receptors used by the circadian system. However, recent studies in mice yielded evidence that the photopigment in mammalian retina used by the circadian system may be a mammalian homolog of the plant blue-light photoreceptor (that is, a cryptochrome based on vitamin B_2 rather than an opsin based on vitamin A).[46-47]

Three major nervous pathways take information to the SCN, as diagrammed in Figure 11. The best known, and perhaps the most important, of these pathways is the *retino-hypothalamic tract* (RHT), a fiber bundle that connects the retina directly (monosynaptically) to the SCN. Usually, but not always, RHT fibers project to the ventrolateral section of the SCN.[22,48-53] Knife cuts that selectively sectioned retinal axons innervating the SCN resulted in loss of entrainment to a light-dark cycle without disrupting free-running rhythms in constant darkness.[54] Electrophysiological recording of neuronal activity indicated that SCN cells respond to photic or electrical stimulation of the retina,[55-59] and electrical stimulation of the distal end of the optic nerve caused behavioral phase shifts equivalent to those produced by light pulses.[60] The cellular mechanism of communication of photic information to the SCN is not well known yet, but it appears that photic input activates SCN cells in nocturnal animals and suppresses the activity of SCN cells in diurnal animals.[59]

Although the neurotransmitter produced by RHT neurons has not been unequivocally identified yet, there is considerable evidence that the transmitter is an excitatory amino acid, particularly glutamate. For instance, optic nerve stimulation causes the release of glutamate and aspartate from hypothalamic explants containing the SCN.[61] MK-801 (a specific antagonist of excitatory-amino-acid receptors) blocks behavioral phase shifts[62] and SCN activation[63] in response to light pulses. MK-801 also blocks phase shifts caused by electrical stimulation of the optic nerve.[60] The administration of either histamine[64] or glutamate[65] to SCN slices *in vitro* causes phase shifts of the clock equivalent to those caused by light pulses (that is, delays during early subjective night and advances during late subjective night). Glutamate administration *in vitro* also affects cellular processes in the SCN (such as induction of the transcription factor Ca^{2+}/cAMP response element-binding protein) in a phase-dependent manner.[66]

The *geniculo-hypothalamic tract* (GHT) is an indirect afferent pathway that relays photic information from the lateral geniculate nucleus of the thalamus

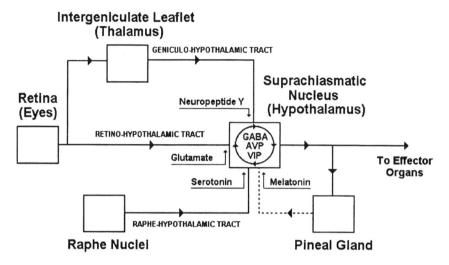

Figure 11. Diagram of the afferent pathways to the suprachiasmatic nucleus in mammals and the neurotransmitters utilized for synaptic transmission.

(a major relay station of the visual system) to the hypothalamus. The GHT projects from the intergeniculate leaflet of the thalamus to the ventrolateral section of the SCN.[67-68] Although the GHT is not required for entrainment to light-dark cycles, it seems to play an auxiliary role.[69-71] The neurotransmitter produced by GHT neurons seems to be the pancreatic polypeptide called "neuropeptide Y" (NPY).[22,72-74] Since microinjection of NPY into the SCN causes phase shifts of circadian rhythms in accordance with the non-photic PRC[75] (in contrast to histamine or glutamate injections, which cause phase shifts in accordance with the photic PRC), it is possible that the GHT serves as the terminal link in the pathway responsible for entrainment by non-photic stimuli. This possibility may apply also to the third major afferent pathway to the SCN: the *raphe-hypothalamic tract*. This tract, which is not technically a tract, but a diffuse path, is part of the serotonergic pathways (that is, nerve fibers that use serotonin — also known as 5-hydroxytryptamine — as a neurotransmitter) that originate in the raphe nuclei and project to various parts of the brain, including the SCN.[76,77] The response of SCN neurons to serotonin administration exhibits circadian rhythmicity, being higher during subjective night than during subjective day.[14] Serotonin agonists cause phase shifts of circadian rhythms according to the non-photic PRC,[78-81] and neurotoxic serotonin depletion was shown to inhibit the non-photic shortening of circadian period caused by running wheel availability.[82] Therefore, the raphe-hypothalamic tract may well be a terminal link in the pathway responsible for entrainment by non-photic stimuli. Against this possibility is the finding that four different serotonin antagonists failed to block phase shifts caused by non-photic stimuli in golden hamsters.[83] Curiously, serotonin precursors and agonists inhibit the response of the circadian clock to photic RHT-mediated stimulation,[84,85] whereas antagonists potentiate the effects of photic stimulation.[86,87] This latter set of observations suggests that activity in the raphe-hypothalamic tract may actually serve as a modulator of the activity in the retino-hypothalamic tract.[88]

In a thorough investigation of neural inputs to the SCN in rats, Moga and Moore identified three additional afferent paths whose function is yet unknown.[89] While the RHT, the GHT, and the path from the raphe nuclei project to the ventrolateral portion of the SCN, the three additional paths have distinct projection sites. The paraventricular portion of the thalamus projects to both the dorsomedial and ventrolateral sections of the SCN as well as to other structures dorsally. The hypothalamus projects mostly to the dorsal section of the SCN as well as to various other brain sites. The limbic telencephalon projects primarily to the dorsal SCN.

other influences

Many drugs not discussed above have occasionally been shown to affect the circadian system. As will be discussed in Chapter 13, antidepressant drugs seem to affect photic sensitivity and free-running period. In addition, morphine injections cause phase shifts of circadian rhythms in mice, although the shifts seem to be due to an acute increase in locomotor activity caused by the drug.[90] Injections of carbachol (an agonist of the neurotransmitter acetylcholine) also produce phase shifts apparently through a non-photic mechanism.[91, 92] Injections of nerve growth factor seem to cause phase shifts similar to those caused by carbachol.[93] Administration of vitamin B_{12} seems to affect photic sensitivity and free-running period in rats, although the mechanism by which this is achieved is not known.[94-96] Perhaps the most interesting effect is that of the drug methamphetamine (an agonist of the neurotransmitter dopamine): systemic administration of methamphetamine in rats previously rendered arrhythmic by SCN lesions induces circadian rhythmicity (with periods in the range of 24.7 to 27.4 hours) in locomotor activity, feeding, drinking, and body temperature.[97] Apparently, there is some nervous structure in the rat that can become rhythmic when stimulated by methamphetamine, although such structure has yet to be identified. Injections of haloperidol (a dopamine antagonist), which presumably block the action of methamphetamine transiently, have been shown to cause phase shifts of the methamphetamine-induced rhythm.[98]

Hormones are another category of chemicals that can affect the operation of the circadian pacemaker. While various hormones have occasionally been shown to affect the operation of the circadian clock (for instance, a slight shortening of circadian period caused by estradiol administration[99,100] and phase shifts caused by thyrotropin-releasing hormone[101]), the hormone *melatonin* — produced mainly by the pineal gland — has received the greatest attention. The pineal gland and the retina of the eyes produce melatonin in practically all vertebrates and, in non-mammalian vertebrates, they do so with circadian rhythmicity *in vitro* — which means that they constitute independent circadian pacemakers.[36] While these pacemakers might be redundant and non-important at the organismal level, their product (melatonin) unquestionably has an important role in the circadian system. For instance, in sparrows, destruction of the SCN abolishes circadian rhythmicity[102] as it does in mammals, but removal of the pineal gland also eliminates rhythmicity in animals with intact SCN maintained in constant darkness (even though rhythmicity persists under a light-dark cycle).[103,104] Daily administration of melatonin restores rhythmicity.[105] In pigeons, the eyes apparently produce enough melatonin to compensate for the loss of the pineal gland, but removal of the eyes after pinealectomy eliminates circadian rhythmicity in constant darkness, and daily infusion of melatonin restores rhythmicity.[106] Similarly, in the quail, removal of the eyes and pineal gland eliminates rhythmicity in animals with intact SCN maintained in constant darkness or under a light-dark cycle.[107]

In mammals, the retina is an independent circadian pacemaker,[108] whereas the pineal is not.[36] Although little is known about melatonin secretion in the mammalian retina, melatonin production by the pineal is known to be under the control of an endogenous component (under the influence of the SCN) as well as of an exogenous component (because melatonin synthesis is acutely inhibited by photic stimulation of the eyes).[109] A robust circadian rhythm of melatonin secretion into the bloodstream can be recorded in humans,[110-114] and melatonin receptors have been identified in the human SCN.[115] Since melatonin secretion is under the control of the SCN, light pulses can phase shift the human rhythm of melatonin secretion (by shifting the circadian clock).[112,113] More significantly, acute administration of melatonin can phase-shift the rhythm of melatonin secretion and other circadian rhythms,[116-120] which indicates that melatonin feedbacks on the SCN.

How important melatonin is in the organization of the circadian system in mammals is not clear. In contrast with the robust results of melatonin administration in humans, research on other mammals have often revealed weak effects of melatonin on circadian rhythms,[121-126] although strong effects have occasionally been reported.[127,128] There is some indication that the infant circadian system in mammals is very sensitive to melatonin but that this sensitivity is greatly reduced in adults.[129-131] It is possible that the adult human circadian system, like that of birds, is more dependent on melatonin than the adult circadian systems of other mammals are. Alternatively, the reliance on the "constant routine" protocol (as opposed to true free-runs) and other confounding procedures in human studies may be responsible for an overestimation of the true effects of melatonin. More rigorous studies in humans are necessary to elucidate this issue. However, regardless of the role of melatonin in the mammalian circadian system, it should be pointed out that melatonin unequivocally plays an important role in mammalian physiology, namely, in the regulation of seasonal reproduction.[132]

efferent pathways

Although the anatomical pathways that connect the SCN to other parts of the brain are relatively well known,[133] there is an embarrassing paucity of information about how the SCN imposes circadian rhythmicity on the various organs of the body. As a matter of fact, it is not even clear whether neural pathways are involved. When researchers created a "hypothalamic island" in the hamster brain by cutting all around the SCN to sever efferent fibers but leaving the SCN in place, circadian rhythmicity was not eliminated.[134] Also, although fetal hypothalamic tissue implanted in the brain of adult animals with SCN lesions restores circadian rhythmicity,[135-140] there is not compelling evidence that nerve outgrowth from the graft is necessary for the restoration of rhythmicity.[141-143] Of course, if neural connections are indeed not necessary, then the physiological action of the SCN must involve a diffusable substance, such as a hormone. To confuse things even further, it has been found that hypothalamic grafts that restore rhythmicity in locomotor activity within a few weeks do not restore rhythmicity of various hormones, at least not within several months.[144] This suggests that endocrine rhythms may require qualitatively different SCN outputs from those required by the locomotor activity rhythm.

One interesting recent finding is that neuronal activity in the ventrolateral preoptic area (VLPO) of the rat is greater during the day than during the night.[145] The VLPO, which receives neural projections from the SCN, is involved in the control of sleep. Since — as previously discussed — all areas of the brain except the SCN are more active during the night than during the day in nocturnal animals, the fact that the VLPO is more active during the day suggests the existence of a direct functional connection between the SCN and the VLPO. Similarly, in golden hamsters, the bed nucleus of the stria terminalis (BNST) was found to be more active during the day, in synchrony with the SCN.[146] Thus, the VLPO and BNST may be components of the SCN's major output pathway. More research on functional SCN efference is definitely needed.

bibliographic notes

1. Page, T. L. (1981). Neural and endocrine control of circadian rhythmicity in invertebrates. In: Aschoff, J. (Ed.). *Handbook of Behavioral Neurobiology, Volume 4: Biological Rhythms.* New York: Plenum, pp.145–172.
2. Dowse, H. B. and Ringo, J. M. (1987). Further evidence that the circadian clock in *Drosophila* is a population of coupled oscillators. *Journal of Biological Rhythms* 2: 65-76.

3. Barrio, R. A., Zhang, L., and Maini, P. K. (1997). Hierarchically coupled ultra-dian oscillators generating robust circadian rhythms. *Bulletin of Mathematical Biology* 59: 517-532.

4. Harrington, M. E., Rahmani, T., and Lee, C. A. (1993). Effects of damage to SCN neurons and efferent pathways on circadian activity rhythms of hamsters. *Brain Research Bulletin* 30: 655-669.

5. Derambure, P. S. and Boulant, J. A. (1994). Circadian thermosensitive characteristics of suprachiasmatic neurons in vitro. *American Journal of Physiology* 266: R1876-R1884.

6. Welsh, D. K., Logothetis, D. E., Meister, M., and Reppert, S. M. (1995). Individual neurons dissociated from rat suprachiasmatic nucleus express independently phased circadian firing rhtyhms. *Neuron* 14: 697-706.

7. Liu, C., Weaver, D. R., Strogatz, S. H., and Reppert, S. M. (1997). Cellular construction of a circadian clock: Period determination in the suprachiasmatic nuclei. *Cell* 91: 855-860.

8. Honma, S., Shirakawa, T., Katsuno, Y., Namihira, M., and Honma, K. (1998). Circadian periods of single suprachiasmatic neurons in rats. *Neuroscience Letters* 250: 157-160.

9. Herzog, E. D., Takahashi, J. S., and Block, G. D. (1998). *Clock* controls circadian period in isolated suprachiasmatic nucleus neurons. *Nature Neuroscience* 1: 708-713.

10. Card, J. P. and Moore, R. Y. (1984). The suprachiasmatic nucleus of the golden hamster: Immunohistochemical analysis of cell and fiber distribution. *Neuroscience* 13: 415-431.

11. van den Pol, A. N. (1986). Gamma-aminobutyrate, gastrin releasing peptide, serotonin, somatostatin, and vasopressin: Ultrastructural immunocytochemical localization in presynaptic axons in the suprachiasmatic nucleus. *Neuroscience* 17: 643-659.

12. Moore, R. Y. and Speh, J. C. (1993). GABA is the principal neurotransmitter of the circadian system. *Neuroscience Letters* 150: 112-116.

13. Shibata, S., Liou, S. Y., and Ueki, S. (1983). Effect of amino acids and monoamines on the neuronal activity of suprachiasmatic nucleus in hypothalamic slice preparations. *Japanese Journal of Pharmacology* 33: 1225-1231.

14. Mason, R. J. (1986). Circadian variation in sensitivity of suprachiasmatic and lateral geniculate neurones to 5-hydroxytryptamine in the rat. *Journal of Physiology* 377: 1-13.

15. Biggs, K. R. and Prosser, R. A. (1998). GABA(B) receptor stimulation phase-shifts the mammalian circadian clock in vitro. *Brain Research* 807: 250-254.

16. Wagner, S., Castel, M., Gainer, H., and Yarom, Y. (1997). GABA in the mammalian suprachiasmatic nucleus and its role in diurnal rhythmicity. *Nature* 387: 598-603.

17. Van Reeth, O., Zhang, Y., Zee, P. C., and Turek, F. W. (1992). Aging alters feedback effects of the activity-rest cycle on the circadian clock. *American Journal of Physiology* 263: R981-R986.

18. Van Reeth, O. and Turek, F. W. (1990). Daily injections of triazolam induce long-term changes in hamster circadian period. *American Journal of Physiology* 259: R514-R520.

19. Wickland, C. and Turek, F. W. (1991). Phase-shifting effect of triazolam on the hamster's circadian rhythm of activity is not mediated by a change in body temperature. *Brain Research* 560: 12-16.

20. Van Reeth, O. and Turek, F. W. (1987). Adaptation of circadian rhythmicity to shift in light-dark cycle accelerated by a benzodiazepine. *American Journal of Physiology* 253: R204-R207.
21. Strecker, G. J., Wuarin, J. P., and Dudek, F. E. (1997). $GABA_A$-mediated local synaptic pathways connect neurons in the rat suprachiasmatic nucleus. *Journal of Neurophysiology* 78: 2217-2220.
22. Smale, L. and Boverhof, J. (1999). The suprachiasmatic nucleus and intergeniculate leaflet of *Arvicanthis niloticus*, a diurnal murid rodent from East Africa. *Journal of Comparative Neurology* 403: 190-208.
23. Okamura, H., Tominaga, K., Ban, Y., Fukuhara, C., Yanaihara, N., Ibata, Y., and Inouye, S. T. (1994). Morphological survey of the suprachiasmatic nucleus in slice culture using roller tube method: 2. Peptidergic neurons. *Acta Histochemica et Cytochemica* 27: 171-179.
24. Murakami, N., Takamure, M., Takahashi, K., Utunomiya, K., Kuroda, H., and Etoh, T. (1991). Long-term cultured neurons from rat suprachiasmatic nucleus retain the capacity for circadian oscillation of vasopressin release. *Brain Research* 545: 347-350.
25. Tominaga, K., Shinohara, K., Otori, Y., Fukuhara, C., and Inouye, S. T. (1992). Circadian rhythms of vasopressin content in the suprachiasmatic nucleus of the rat. *NeuroReport* 3: 809-812.
26. Gerkema, M. P., Van der Zee, E. A., and Feitsma, L. E. (1994). Expression of circadian rhythmicity correlates with the number of arginine-vasopressin-immunoreactive cells in the suprachiasmatic nucleus of common voles, *Microtus arvalis*. *Brain Research* 639: 93-101.
27. Shibata, S. and Moore, R. Y. (1988). Neuropeptide Y and vasopressin effects on rat suprachiasmatic nucleus neurons in vitro. *Journal of Biological Rhythms* 3: 265-276.
28. Liou, S. Y. and Albers, H. E. (1989). Single unit response of suprachiasmatic neurons to arginine vasopressin (AVP) is mediated by a V1-like receptor in the hamster. *Brain Research* 477: 336-343.
29. Schilling, J. and Nurnberger, F. (1998). Dynamic changes in the immunoreactivity of neuropeptide systems of the suprachiasmatic nuclei in golden hamster during the sleep–wake cycle. *Cell and Tissue Research* 294: 233-241.
30. Card, J. P., Brecha, N., Kasten, H. J., and Moore, R. Y. (1981). Immunocytochemical localization of vasoactive intestinal polypeptide-containing cells and processes in the suprachiasmatic nucleus of the rat: Light and electron microscopic analysis. *Journal of Neuroscience* 1: 1289-1303.
31. Martin, J. L., Dietl, M. M., Hof, P. R., Palacios, J. M., and Magistretti, P. J. (1987). Autoradiographic mapping of [mono[^{125}I]iodo-Tyr^{10}, Met-O^{17}]vasoactive intestinal peptide binding sites in the rat brain. *Neuroscience* 23: 539-565.
32. [^{125}I]Vasoactive intestinal peptide binding in rodent suprachiasmatic nucleus: Developmental and circadian studies. *Brain Research* 605: 271-279.
33. Piggins, H. D., Antle, M. C., and Rusak, B. (1995). Neuropeptides phase shift the mammalian circadian pacemaker. *Journal of Neuroscience* 15: 5612-5622.
34. Cagampang, F. R. A., Okamura, H., and Inouye, S. T. (1994). Circadian rhythms of norepinephrine in the rat suprachiasmatic nucleus. *Neuroscience Letters* 173: 185-188.
35. Cahill, A. L. and Ehret, C. F. (1982). Alpha-methyl-p-tyrosine shifts circadian temperature rhythm. *American Journal of Physiology* 243: R218-R222.

36. Menaker, M., Moreira, L. F., and Tosini, G. (1997). Evolution of circadian organization in vertebrates. *Brazilian Journal of Medical and Biological Research* 30: 305-313.
37. Campbell, S. S. and Murphy, P. J. (1998). Extraocular circadian phototransduction in humans. *Science* 279: 396-399.
38. Richter, C. P. (1970). Dependence of successful mating in rats on functioning of the 24-hour clocks of the male and female. *Communications in Behavioral Biology* 5: 1-5.
39. Zucker, I. (1971). Light-dark rhythms in rat eating and drinking behavior. *Physiology and Behavior* 6: 115-126.
40. Refinetti, R., Nelson, D.E., and Menaker, M. (1992). Social stimuli fail to act as entraining agents of circadian rhythms in the golden hamster. *Journal of Comparative Physiology A* 170: 181-187.
41. Foster, R. G., Provencio, I., Hudson, D., Fiske, S., De Grip, W., and Menaker, M. (1991). Circadian photoreception in the retinally degenerate mouse (*rd/rd*). *Journal of Comparative Physiology A* 169: 39-50.
42. Lovegrove, B. G., Heldmeier, G., and Ruf, T. (1993). Circadian activity rhythms in colonies of "blind" molerats, *Cryptomys damarensis* (Bathyergidae). *South African Journal of Zoology* 28: 46-55.
43. Lovegrove, B. G. and Papenfus, M. E. (1995). Circadian activity rhythms in the solitary cape mole rat (*Georychus capensis*: Bathyergidae) with some evidence of splitting. *Physiology and Behavior* 58: 679-685.
44. McGuire, R. A., Rand, W. M., and Wurtman, R. J. (1973). Entrainment of the body temeperature rhythm in rats: Effect of color and intensity of environmental light. *Science* 181: 956-957.
45. Takahashi, J. S., DecOursey, P. J., Bauman, L., and Menaker, M. (1984). Spectral sensitivity of a novel photoreceptive system mediating entrainment of mammalian circadian rhythms. *Nature* 308: 186-188.
46. Miyamoto, Y. and Sancar, A. (1998). Vitamin B_2-based blue-light photoreceptors in the retinohypothalamic tract as the photoactive pigments for setting the circadian clock in mammals. *Proceedings of the National Academy of Sciences* 95: 6097-6102.
47. Thresher, R. J., Vitaterna, M. H., Miyamoto, Y., Kazantsev, A., Hsu, D. S., Petit, C., Selby, C. P., Dawut, L., Smithies, O., Takahashi, J. S., and Sancar, A. (1998). Role of mouse cryptochrome blue-light photoreceptor in circadian photoresponses. *Science* 282: 1490-1494.
48. Moore, R. Y. and Lenn, N. J. (1972). A retinohypothalamic projection in the rat. *Journal of Comparative Neurology* 146: 1-148.
49. Moore, R. Y. (1973). Retinohypothalamic projection in mammals: A comparative study. *Brain Research* 49: 403-409.
50. Kudo, M., Yamamoto, M., and Nakamura, Y. (1991). Suprachiasmatic nucleus and retinohypothalamic projections in moles. *Brain Behavior and Evolution* 38: 332-338.
51. Stopa, E. G., Johnson, J. K., Friedman, D. I., Ryer, H. I., Reidy, J., Kuo-LeBlanc, V., and Albers, H. E. (1995). Neuropeptide Y receptor distribution and regulation in the suprachiasmatic nucleus of the Syrian hamster (*Mesocricetus auratus*). *Peptide Research* 8: 95-100.
52. Murakami, D. M. and Fuller, C. A. (1990). The retinohypothalamic projection and oxidative metabolism in the suprachiasmatic nucleus of primates and tree shrews. *Brain Behavior and Evolution* 35: 302-312.

53. Sato, T. and Kawamura, H. (1984). Effects of bilateral suprachiasmatic nucleus lesions on the circadian rhythms in a diurnal rodent, the Siberian chipmunk (*Eutamias sibiricus*). *Journal of Comparative Physiology A* 155: 745-752.

54. Johnson, R. F., Moore, R. Y., and Morin, L. P. (1988). Loss of entrainment and anatomical plasticity after lesions of the hamster retinohypothalamic tract. *Brain Research* 460: 297-313.

55. Roig, J. A., Granados-Fuentes, D., and Aguilar-Roblero, R. (1997). Neuronal subpopulations in the suprachiasmatic nuclei based on their response to retinal and intergeniculate leaflet stimulation. *NeuroReport* 8: 885-890.

56. Senseman, D. M. and Rea, M. A. (1994). Fast multisite optical recording of mono- and polysynaptic activity in the hamster suprachiasmatic nucleus evoked by retinohypothalamic tract stimulation. *Neuroimage* 1: 247-263.

57. Meijer, J. H., Rusak, B., and Gänshirt, G. (1992). The relation between light-induced discharge in the suprachiasmatic nucleus and phase shifts of hamster circadian rhythms. *Brain Research* 598: 257-263.

58. Meijer, J. H., Watanabe, K., Schaap, J., Albus, H., and Détári, L. (1998). Light responsiveness of the suprachiasmatic nucleus: Long-term multiunit and single-unit recordings in freely moving rats. *Journal of Neuroscience* 18: 9078-9087.

59. Jiao, Y. Y., Lee, T. M., and Rusak, B. (1999). Photic responses of suprachiasmatic area neurons in diurnal degus (*Octodon degus*) and nocturnal rats (*Rattus norvegicus*). *Brain Research* 817: 93-103.

60. de Vries, M. J., Treep, J. A., de Pauw, E. S. D., and Meijer, J. H. (1994). The effects of electrical stimulation of the optic nerves and anterior optic chiasm on the circadian activity rhythm of the Syrian hamster: Involvement of excitatory amino acids. *Brain Research* 642: 206-212.

61. Liou, S. Y., Shibata, S., Iwasaki, K., and Ueki, S. (1986). Optic nerve stimulation-induced increase of release of ^{3}H-glutamate and ^{3}H-aspartate but not ^{3}H-GABA from the suprachiasmatic nucleus in slices of the rat hypothalamus. *Brain Research Bulletin* 16: 527-531.

62. Colwell, C. S., Ralph, M. R., and Menaker, M. (1990). Do NMDA receptors mediate the effects of light on circadian behavior? *Brain Research* 523: 117-120.

63. Edelstein, K. and Amir, S. (1998). Glutamatergic antagonists do not attenuate light-induced Fos protein in rat intergeniculate leaflet. *Brain Research* 810: 264-268.

64. Cote, N. K. and Harrington, M. E. (1993). Histamine phase shifts the circadian clock in a manner similar to light. *Brain Research* 613: 149-151.

65. Shirakawa, T. and Moore, R. Y. (1994). Glutamate shifts the phase of the circadian neuronal firing rhythm in the rat suprachiasmatic nucleus in vivo. *Neuroscience Letters* 178: 47-50.

66. von Gall, C., Duffield, G. E., Hastings, M. H., Kopp, M. D. A., Dehghani, F., Korf, H. W., and Stehle, J. H. (1998). CREB in the mouse SCN: A molecular interface coding the phase-adjusting stimuli light, glutamate, PACAP, and melatonin for clockwork access. *Journal of Neuroscience* 18: 10389-10397.

67. Card, J. P. and Moore, R. Y. (1982). Ventral lateral geniculate nucleus efferents to the rat suprachiasmatic nucleus exhibit avian pancreatic polypeptide-like immunoreactivity. *Journal of Comparative Neurology* 206: 390-396.

68. Pickard, G. E. (1982). The afferent connections of the suprachiasmatic nucleus of the golden hamster with emphasis on the retinohypothalamic projection. *Journal of Comparative Neurology* 211: 65-83.

69. Harrington, M. E. and Rusak, B. (1988). Ablation of the geniculo-hypothalamic tract alters circadian activity rhythms of hamsters housed under constant light. *Physiology and Behavior* 42: 183-189.

70. Rusak, B., Meijer, J. H., and Harrington, M. E. (1989). Hamster circadian rhythms are phase-shifted by electrical stimulation of the geniculo-hypothalamic tract. *Brain Research* 493: 283- .

71. Edelstein, K. and Amir, S. (1999). The role of the intergeniculate leaflet in entrainment of circadian rhythms to a skeleton photoperiod. *Journal of Neuroscience* 19: 372-380.

72. Harrington, M. E., Nance, D. M., and Rusak, B. (1985). Neuropeptide Y immunoreactivity in the hamster geniculo-suprachiasmatic tract. *Brain Research Bulletin* 15: 465-472.

73. Harrington, M. E., Nance, D. M., and Rusak, B. (1987). Double-labeling of neuropeptide Y-immunoreactive neurons which project from the geniculate to the suprachiasmatic nuclei. *Brain Research* 410: 275-282.

74. Card, J. P. and Moore, R. Y. (1989). Organization of lateral geniculate-hypothalamic connections in the rat. *Journal of Comparative Neurology* 284: 135-147.

75. Huhman, K. L. and Albers, H. E. (1994). Neuropeptide Y microinjected into the suprachiasmatic region phase shifts circadian rhythms in constant darkness. *Peptides* 15: 1475-1478.

76. Aghajanian, G. K., Bloom, F. E., and Sheard, M. H. (1969). Electron microscopy of degeneration within the serotonin pathway of rat brain. *Brain Research* 13: 266-273.

77. Moore, R. Y., Halaris, A. E., and Jones, B. E. (1978). Serotonin neurons of the midbrain raphe: Ascending projections. *Journal of Comparative Neurology* 180: 417-438.

78. Cutrera, R. A., Saboureau, M., and Pévet, P. (1996). Phase-shifting effect of 8-OH-DPAT, a $5\text{-}HT_{1A}/5\text{-}HT_7$ receptor agonist, on locomotor activity in golden hamster in constant darkness. *Neuroscience Letters* 210: 1-4.

79. Bobrzynska, K. J., Godfrey, M. H., and Mrosovsky, N. (1996). Serotonergic stimulation and nonphotic phase-shifting in hamsters. *Physiology and Behavior* 59: 221-230.

80. Prosser, R. A., Macdonald, E. S., and Heller, H. C. (1994). C-*fos* mRNA in the suprachiasmatic nuclei in vitro shows a circadian rhythm and responds to a serotonergic agonist. *Molecular Brain Research* 25: 151-156.

81. Mintz, E. M., Gillespie, C. F., Marvel, C. L., Huhman, K. L., and Albers, H. E. (1997). Serotonergic regulation of circadian rhythms in Syrian hamsters. *Neuroscience* 79: 563-569.

82. Mistlberger, R. E., Bossert, J. M., Holmes, M. M., and Marchant, E. G. (1998). Serotonin and feedback effects of behavioral activity on circadian rhythms in mice. *Behavioural Brain Research* 96: 93-99.

83. Antle, M. C., Marchant, E. G., Niel, L., and Mistlberger, R. E. (1998). Serotonin antagonists do not attenuate activity-induced phase shifts of circadian rhythms in the Syrian hamster. *Brain Research* 813: 139-149.

84. Glass, J. D., Selim, M., Srkalovic, G., and Rea, M. A. (1995). Tryptophan loading modulates light-induced responses in the mammalian circadian system. *Journal of Biological Rhythms* 10: 80-90.

85. Selim, M., Glass, J. D., Hauser, U. E., and Rea, M. A. (1993). Serotonergic inhibition of light-induced *fos* protein expression and extracellular glutamate in the suprachiasmatic nuclei. *Brain Research* 621: 181-188.

86. Rea, M. A., Barrera, J., Glass, J. D., and Gannon, R. L. (1995). Serotonergic potentiation of photic phase shifts of the circadian activity rhythm. *NeuroReport* 6: 1289-1292.

87. Mistlberger, R. E. and Antle, M. C. (1998). Behavioral inhibition of light-induced circadian phase resetting is phase and serotonin dependent. *Brain Research* 786: 31-38.

88. Moriya, T., Yoshinobu, Y., Ikeda, M., Yokota, S., Akiyama, M., and Shibata, S. (1998). Potentiating action of MKC-242, a selective 5-HT$_{1A}$ receptor agonist, on the photic entrainment of the circadian activity rhythm in hamsters. *British Journal of Pharmacology* 125: 1281-1287.

89. Moga, M. M. and Moore, R. Y. (1997). Organization of neural inputs to the suprachiasmatic nucleus in the rat. *Journal of Comparative Neurology* 389: 508-534.

90. Marchant, E. G. and Mistlberger, R. E. (1995). Morphine phase-shifts circadian rhythms in mice: Role of behavioral activation. *NeuroReport* 7: 209-212.

91. Wee, B. E. F., Anderson, K. D., Kouchis, N. S., and Turek, F. W. (1992). Administration of carbachol into the lateral ventricle and suprachiasmatic nucleus (SCN) produces dose-dependent phase shifts in the circadian rhythm of locomotor activity. *Neuroscience Letters* 137: 211-215.

92. Colwell, C. S., Kaufman, C. M., and Menaker, M. (1993). Phase-shifting mechanisms in the mammalian circadian system: New light on the carbachol paradox. *Journal of Neuroscience* 13: 1454-1459.

93. Bina, K. G. and Rusak, B. (1996). Nerve growth factor phase shifts circadian activity rhythms in Syrian hamsters. *Neuroscience Letters* 206: 97-100.

94. Ikeda, M., Honda, K., and Inoué, S. (1996). Vitamin B$_{12}$ amplifies circadian phase shifts induced by a light pulse in rats. *Experientia* 52: 691-694.

95. Takahashi, Y., Usui, S., and Honda, Y. (1992). Effect of vitamin B$_{12}$ (mecobalamin) on the free-running period of rat circadian behavioral rhythm. *Japanese Journal of Psychiatry and Neurology* 46: 222-224.

96. Tsujimaru, S., Egami, H., Honma, G., Ida, Y., Musaka, H., and Nakazawa, Y. (1992). Effects of vitamin B$_{12}$ on the period of free-running rhythm in rats. *Japanese Journal of Psychiatry and Neurology* 46: 225-226.

97. Honma, S., Honma, K., Shirakawa, T., and Hiroshige, T. (1988). Rhythms in behavior, body temperature and plasma corticosterone in SCN lesioned rats given methamphetamine. *Physiology and Behavior* 44: 247-255.

98. Honma, S. and Honma, K. (1995). Phase-dependent phase shift of methamphetamine-induced circadian rhythm by haloperidol in SCN-lesioned rats. *Brain Research* 674: 283-290.

99. Morin, L. P., Fitzgerald, K. M., and Zucker, I. (1977). Estradiol shortens the period of hamster circadian rhythms. *Science* 196: 305-307.

100. Takahashi, J. S. and Menaker, M. (1980). Interaction of estradiol and progesterone: Effects on circadian locomotor rhythm of female golden hamsters. *American Journal of Physiology* 239: R497-R504.

101. Gary, K. A., Sollars, P. J., Lexow, N., Winokur, A., and Pickard, G. E. (1996). Thyrotropin-releasing hormone phase shifts circadian rhythms in hamsters. *NeuroReport* 7: 1631-1634.

102. Takahashi, J. S. and Menaker, M. (1982). Role of the suprachiasmatic nuclei in the circadian system of the house sparrow, *Passer domesticus*. *Journal of Neuroscience* 2: 815-828.

103. Gaston, S. and Menaker, M. (1968). Pineal function: The biological clock in the sparrow? *Science* 160: 1125-1127.
104. Binkley, S., Kluth, E., and Menaker, M. (1971). Pineal function in sparrows: Circadian rhythms and body temperature. *Science* 174: 311-314.
105. Heigl, S. and Gwinner, E. (1994). Periodic melatonin in the drinking water synchronizes circadian rhythms in sparrows. *Naturwissenschaften* 81: 83-85.
106. Chabot, C. C. and Menaker, M. (1992). Effects of physiological cycles of infused melatonin on circadian rhythmicity in pigeons. *Journal of Comparative Physiology A* 170: 615-622.
107. Underwood, H. (1994). The circadian rhythm of thermoregulation in Japanese quail. I. Role of the eyes and pineal. *Journal of Comparative Physiology A* 175: 639-653.
108. Tosini, G. and Menaker, M. (1996). Circadian rhythms in cultured mammalian retina. *Science* 272: 419-421.
109. Lewy, A. J., Wehr, T. A., Goodwin, F. K., Newsome, D. A., and Markey, S. P. (1980). Light supresses melatonin secretion in humans. *Science* 210: 1267-1269.
110. Kennaway, D. J. and van Dorp, C. F. (1991). Free-running rhythms of melatonin, cortisol, electrolytes, and sleep in humans in Antarctica. *American Journal of Physiology* 260: R1137-R1144.
111. Honma, K. I., Honma, S., Nakamura, K., Sasaki, M., Endo, T., and Takahashi, T. (1995). Differential effects of bright light and social cues on reentrainment of human circadian rhythms. *American Journal of Physiology* 268: R528-R535.
112. Buresová, M., Dvoráková, M., Zvolsky, P., and Illnerová, H. (1991). Early morning bright light phase advances the human circadian pacemaker within one day. *Neuroscience Letters* 121: 47-50.
113. Shanahan, T. L. and Czeisler, C. A. (1991). Light exposure induces equivalent phase shifts of the endogenous circadian rhythms of circulating plasma melatonin and core body temperature in men. *Journal of Clinical Endocrinology and Metabolism* 73: 227-235.
114. Matsumoto, M., Sack, R. L., Blood, M. L., and Lewy, A. J. (1997). The amplitude of endogenous melatonin production is not affected by melatonin treatment in humans. *Journal of Pineal Research* 22: 42-44.
115. Weaver, D. R. and Reppert, S. M. (1996). The Mel_{1a} melatonin receptor gene is expressed in human suprachiasmatic nuclei. *NeuroReport* 8: 109-112.
116. Sack, R. L., Lewy, A. J., Blood, M. L., Stevenson, J., and Keith, L. D. (1991). Melatonin administration to blind people: Phase advances and entrainment. *Journal of Biological Rhythms* 6: 249-261.
117. Lewy, A. J., Ahmed, S., Jackson, J. M. L., and Sack, R. L. (1992). Melatonin shifts human circadian rhythms according to a phase–response curve. *Chronobiology International* 9: 380-392.
118. Deacon, S., English, J., and Arendt, J. (1994). Acute phase-shifting effects of melatonin associated with suppression of core body temperature in humans. *Neuroscience Letters* 178: 32-34.
119. Lewy, A. J., Ahmed, S., and Sack, R. L. (1996). Phase shifting the human circadian clock using melatonin. *Behavioural Brain Research* 73: 131-134.
120. Attenburrow, M. E. J., Dowling, B. A., Sargent, P. A., Sharpley, A. L., and Cowen, P. J. (1995). Melatonin phase advances circadian rhythm. *Psychopharmacology* 121: 503-505.
121. Ruby, N. F., Ibuka, N., Barnes, B. M., and Zucker, I. (1989). Suprachiasmatic nuclei influence torpor and circadian temperature rhythms in hamsters. *American Journal of Physiology* 257: R210-R215.

122. Hastings, M. H., Mead, S. M., Vindlacheruvu, R. R., Ebling, F. J. P., Maywood, E. S., and Grosse, J. (1992). Non-photic phase shifting of the circadian activity rhythm of Syrian hamsters: The relative potency of arousal and melatonin. *Brain Research* 591: 20-26.

123. Cheung, P. W. and McCormack, C. E. (1982). Failure of pinealectomy or melatonin to alter circadian activity rhythm of the rat. *American Journal of Physiology* 242: R261-R264.

124. Chesworth, M. J., Cassone, V. M., and Armstrong, S. M. (1987). Effects of daily melatonin injections on activity rhythms of rats in constant light. *American Journal of Physiology* 253: R101-R107.

125. Warren, W. S. and Cassone, V. M. (1995). The pineal gland: Photoreception and coupling of behavioral, metabolic, and cardiovascular circadian outputs. *Journal of Biological Rhythms* 10: 64-79.

126. Bobbert, A. C. and Riethoven, J. J. (1991). Feedback in the rabbit's central circadian system, revealed by the changes in its free-running food intake pattern induced by blinding, cervical sympathectomy, pinealectomy, and melatonon administration. *Journal of Biological Rhythms* 6: 263-278.

127. Kirsch, R., Belgnaoui, S., Gourmelen, S., and Pévet, P. (1993). Daily melatonin infusion entrains free-running activity in Syrian and Siberian hamsters. In: Wetterberg, L. (Ed.). *Light and Biological Rhythms in Man*. New York: Pergamon, pp. 107-120.

128. Sumová, A. and Illnerová, H. (1996). Melatonin instantaneously resets intrinsic circadian rhythmicity in the rat suprachiasmatic nucleus. *Neuroscience Letters* 218: 181-184.

129. Davis, F. C. and Mannion, J. (1988). Entrainment of hamster pup circadian rhythms by prenatal melatonin injections to the mother. *American Journal of Physiology* 255: R439-R448.

130. Duffield, G. E., Hastings, M. H., and Ebling, F. J. P. (1998). Investigation into the regulation of the circadian system by dopamine and melatonin in the adult Siberian hamster (*Phodopus sungorus*). *Journal of Neuroendocrinology* 10: 871-884.

131. Grosse, J. and Davis, F. C. (1998). Melatonin entrains the restored circadian activity rhythms of Syrian hamsters bearing fetal suprachiasmatic nucleus grafts. *Journal of Neuroscience* 18: 8032-8037.

132. Karsch, F. J., Woodfill, C. J. I., Malpaux, B., Robinson, J. E., and Wayne, N. L. (1991). Melatonin and mammalian photoperiodism: Synchronization of annual reproductive cycles. In: Klein, D. C., Moore, R. Y., and Reppert, S. M. (Eds.). *Suprachiasmatic Nucleus: The Mind's Clock*. New York: Oxford University Press, pp. 217-232.

133. Watts, A. G. (1991). The efferent projections of the suprachiasmatic nucleus: Anatomical insights into the control of circadian rhythms. In: Klein, D. C., Moore, R. Y., and Reppert, S. M. (Eds.). *Suprachiasmatic Nucleus: The Mind's Clock*. New York: Oxford University Press, pp. 77-106.

134. Hakim, H., DeBernardo, A. P., and Silver, R. (1991). Circadian locomotor rhythms, but not photoperiodic responses, survive surgical isolation of the SCN in hamsters. *Journal of Biological Rhythms* 6: 97-113.

135. DeCoursey, P. J. and Buggy, J. (1988). Restoration of circadian locomotor activity in arrhythmic hamsters by fetal SCN transplants. *Comparative Endocrinology* 4: 49-54.

136. Ralph, M. R., Foster, R. G., Davis, F. C., and Menaker, M. (1990). Transplanted suprachiasmatic nucleus determines circadian period. *Science* 247: 975-978.

137. Saitoh, Y., Matsui, Y., Nihonmatsu, I., and Kawamura, H. (1991). Cross-species transplantation of the suprachiasmatic nuclei from rats to Siberian chipmunks with suprachiasmatic lesions. *Neuroscience Letters* 123: 77-81.

138. Romero, M. T., Lehman, M. N., and Silver, R. (1993). Age of donor influences ability of suprachiasmatic nucleus grafts to restore circadian rhythmicity. *Developmental Brain Research* 71: 45-52.

139. Le Sauter, J. and Silver, R. (1994). Suprachiasmatic nucleus lesions abolish and fetal grafts restore circadian gnawing rhythms in hamsters. *Restorative Neurology and Neuroscience* 6: 135-143.

140. Sollars, P. J., Kimble, D. P., and Pickard, G. E. (1995). Restoration of circadian behavior by anterior hypothalamic heterografts. *Journal of Neuroscience* 15: 2109-2122.

141. Lehman, M. N., LeSauter, J., Kim, C., Berriman, S. J., Tresco, P. A., and Silver, R. (1995). How do fetal grafts of the suprachiasmatic nucleus communicate with the host brain? *Cell Transplantation* 4: 75-81.

142. LeSauter, J., Romero, P., Cascio, M., and Silver, R. (1997). Attachment site of grafted SCN influences precision of restored circadian rhythm. *Journal of Biological Rhythms* 12: 327-338.

143. Lehman, M. N., LeSauter, J., and Silver, R. (1998). Fiber outgrowth from anterior hypothalamic and cortical xenografts in the third ventricle. *Journal of Comparative Neurology* 391: 133-145.

144. Meyer-Bernstein, E. L., Jetton, A. E., Matsumoto, S. I., Markuns, J. F., Lehman, M. N., and Bittman, E. L. (1999). Effects of suprachiasmatic transplants on circadian rhythms of neuroendocrine function in golden hamsters. *Endocrinology* 140: 207-218.

145. Novak, C. M. and Nunez, A. A. (1998). Daily rhythms in Fos activity in the rat ventrolateral preoptic area and midline thalamic nuclei. *American Journal of Physiology* 275: R1620-R1626.

146. Yamazaki, S., Kerbeshian, M. C., Hocker, C. G., Block, G. D., and Menaker, M. (1998). Rhythmic properties of the hamster suprachiasmatic nucleus *in vivo*. *Journal of Neuroscience* 18: 10709-10723.

chapter ten

---

molecular structure of the clock

> The molecular mechanisms (gene activation, transcription, and translation) of circadian rhythms are described. Despite significant interspecies differences, there is a remarkable similarity between the circadian systems of all life forms.

Molecular biology is the star science at the threshold of the 21st century. The amazing applications of knowledge generated by molecular biologists are in the newspaper headlines and movie pictures. Steven Spielberg's two motion pictures based on Michael Crichton's book *Jurassic Park*,[1] added fantasy to but did not create the fact that DNA (the nucleic acid that carries genetic information) can be retrieved from the remains of life forms that existed thousands or even millions of years in the past. Dolly the ewe made the headlines as the first demonstration that a fully-functional mammal can be cloned from a single adult cell.[2] The rate of genetic mutation has been used as a "molecular clock" to measure archeological time and to determine branching of species during evolution.[3] DNA fingerprinting was used to prove that the stain in Ms. Lewinsky's dress was semen from President Clinton.[4] Many more headlines can be anticipated soon after completion of the Human Genome Project, which is cataloging every gene in the human genome.[5]

Just as impressive as the technical achievements of molecular biology are its philosophical implications. Almost as an empirical refutation of the arguments of philosophers of science during the second half of the 20th century, molecular biology provides the physico-chemical bases of biological processes. Behavioral characteristics of whole organisms are practically explained by the chemical mechanisms of gene activation and transcription. It is almost as if proof had finally been provided for the claims of logical positivism (in the first half of the 20th century) that complex phenomena can be reduced to simple elements,[6] that all sciences can ultimately be reduced to the physical sciences.[7] A natural expansion of this is the solution of the old mind-body dichotomy by the assertion that the mind is an expression of the operation of the brain cells.[8,9] This "radical materialism" may be a little bit too much for most readers, but it will certainly become more acceptable to future generations as progress in molecular biology proceeds.

Before examining what molecular biology has to say about circadian rhythms, it is important to point out that the relative simplicity of the phenomena described here is the result of a careful presentation of the most significant features and does not reflect the complexity of actual research in the field. Because of the rapid advances in scientific research, basic college education cannot possibly prepare the average citizen to fully understand research conducted during his or her lifetime. As a reminder of this simple fact, I reproduce here a short paragraph from the Methods section of a representative article dealing with the molecular aspects of circadian rhythms:[10]

> *Bgl* II fragments comprising 5.3 kbp of the *frq* genomic clone (containing the N-terminal 893 codons of FRQ, the 5′ untranslated region, >1200 bp of 5′ nontranscribed DNA, and DNA encoding the 5′ half of the small *frq* transcript) were replaced with a 1.5-kbp HygB^R cassette under control of the *Aspergillus trpC* promoter. A *Spe* I fragment containing the entire HygB^R-interrupted version of *frq* was then exchanged into the Bml^R construct pCRM101 to yield pCMW1. This final construct thus contains two dominant selectable markers, one of which (Bml^R) is lost upon homologous reciprocal integration of this modified *frq* into the genome in place of the resident copy of *frq*.

If the passage was incomprehensible to you, rest assured that it is also incomprehensible to more than 99.9% of the world population. However, let us not allow it to spoil our day. Let us proceed.

As you might expect, more is known about the molecular biology of simpler organisms, such as fungi and fruit flies, than about more complex organisms, such as mammals. Nevertheless, some molecular processes that take place in the suprachiasmatic nucleus are known. The c-fos gene and its product, the Fos protein, are a good example. The *fos* gene is an "immediate early gene," which means that, in the process of carrying out some cellular function, *fos* activation is the first step in a series of reactions. The *fos* gene, like most genes (which are located in the chromosomes in the nucleus of the cell), is normally inactive. When it is activated, it transcribes its doublestranded DNA sequence into a single-stranded RNA sequence (the messenger RNA, or mRNA). The mRNA molecule leaves the cell nucleus and is translated into the Fos protein. The Fos protein then re-enters the nucleus, where it regulates the transcription of other, "late response" genes. Well, it turns out that photic stimulation of the eyes in whole animals induces *fos* expression in the SCN cells — and this happens only when light is presented during the sensitive zone of the PRC.[11-13] Interestingly, *fos* expression is mediated by CREB (cyclic adenosine monophosphate response element binding protein), and light pulses that cause phase shifts of circadian rhythms were found to induce phosphorilation of CREB in SCN cells.[14-15]

Although *fos* induction by photic stimulation has been replicated many times since its first description in 1990,[16-20] it is still not clear whether Fos is equally involved in phase delays and phase advances of the clock. Phase-shifts caused by non-photic stimuli do not seem to involve *fos* induction.[17,21-23] A recent study provided evidence that *fos* induction by light (during subjective night) is restricted to the ventrolateral region of the SCN.[24] You may recall from Chapter 9 that this is the area where the fibers from the retino-hypothalamic tract usually arrive. While the dorsomedial region does not seem to respond to photic stimulation, it does show an endogenous rhythm of *fos* induction in constant darkness.[24]

What about fungi and fruit flies? In Chapter 5, it was mentioned that specific single-gene mutations had been identified in the fruit fly *Drosophila melanogaster*,[25] and in the bread mold *Neurospora crassa*,[26] early in the 1970s. In the fruit fly, the gene — called *per*, for *period* — is located in the X chromosome and exhibits multiple mutant alleles. The first three alleles that were identified shortened (per^S, $\tau \approx 19$ hours), lengthened (per^L, $\tau \approx 29$ hours), or eliminated (per^0) circadian rhythmicity of activity. Other alleles were identified later.[27] The *per* gene is, of course, present in cells of all organs of the fly. Curiously, it is activated in most organs, so that PER (the protein produced by the *per* gene) is produced rhythmically almost everywhere in the fly's body, and this rhythm can be phase-shifted by light pulses.[28] However, only *per* expression in nerve cells and glia seems to be necessary for circadian rhythmicity of locomotor activity.[29]

The *per* gene in the fruit fly is expressed in a circadian manner, and fluctuations in *per* mRNA abundance are influenced by its own translation product (the PER protein).[30] Accordingly, transient experimenter-induced increases in PER production cause phase shifts of the activity rhythm.[31] The exciting aspect of this is that the *per*-PER relationship constitutes the type of feedback loop that we briefly discussed in Chapter 5 as the basic mechanism of biological timing. As diagrammed in Figure 12A, a biochemical process (*per* transcription) is regulated by its product (the PER protein) through a simple loop. Of course, there are thousands or millions of feedback loops in the body, but very few of them involve genes that are known to affect the circadian system. The idea that the *per*-PER loop might be *the* mechanism responsible for the generation of circadian rhythmicity is certainly very exciting. Dr. Jaap van Pelt, in Holland, developed a mathematical model based on a generic mRNA-protein loop such as this one that generates robust circadian oscillation with realistic entrainment and phase-response properties.[32]

In 1994, another *Drosophila* gene involved in the control of circadian rhythmicity — *tim*, short for *timeless*, located in chromosome 2 — was

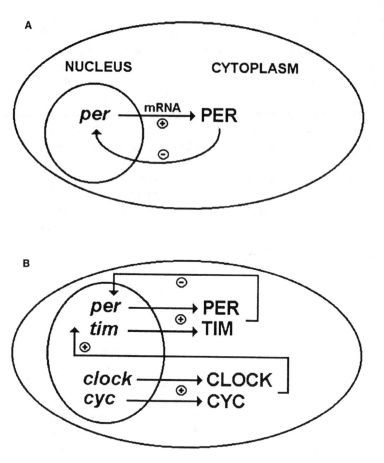

Figure 12. Diagrams of circadian gene loops: (A) a simple *per*-PER negative feedback loop; (B) positive and negative feedback loops involving four genes and their proteins.

identified.[33] Similarly to *per*, *tim* RNA accumulates rhythmically, and the rhythmicity is dependent on the presence of both TIM and PER.[34] Indeed, TIM binds to PER *in vitro*,[35] and TIM degradation correlates with behavioral phase-shifts in response to light pulses.[36] Light-induced disruption of the TIM-PER complex *in vivo* seems to be a component of the process that differentiates phase shifts into phase advances and phase delays.[37] So, the *per*-PER loop is not a one-gene loop. Still, a loop involving a few genes and a few proteins could very well be the biochemical mechanism of circadian rhythmicity.

In the bread mold, over a dozen clock mutations have been isolated, eight of them at the *frq* (short for *frequency*) locus. Similarly to *per* in the fruit fly, the *frq* gene in the bread mold encodes a product (the FRQ protein) that negatively feeds back on the gene to regulate its transcription on a circadian time scale.[38] Again, whether this feedback loop is the fundamental process responsible for the generation of circadian rhythmicity has yet to be determined. The abundance of *frq* mRNA increases in response to photic stimulation, and the magnitude of this increase correlates with the magnitude of phase shifts of the "activity" rhythm (production of asexual spores).[39] Two additional genes, called white collar (*wc-1* and *wc-2*), also seem to be involved, as the proteins WC-1 and WC-2 are necessary for the manifestation of the circadian rhythm in *frq* expression.[40]

Has similar research been conducted in mammals? Yes. Two single-gene mutations in mammals were mentioned in Chapter 5: the *tau* gene in the golden hamster[41] and the *Clock* gene in the mouse.[42] We have already discussed the physiological action of the *tau* gene, and very little is known about its molecular mechanism. The *Clock* gene was isolated in 1994 by a team of researchers led by Dr. Joseph Takahashi at Northwestern University, in Illinois. Many years earlier, Takahashi had been a graduate student of Dr. Michael Menaker's at the University of Oregon. When Menaker identified the *tau* mutation in 1989, Takahashi (who graduated in 1981) was very excited about it but, of course, he wished that *he* had been the one to identify the first mammalian gene involved in circadian rhythmicity. More importantly, the *tau* gene was not very useful to molecular biologists, as the molecular biology of the hamster was not (and still is not) well known. A few years later, however, Takahashi, Menaker, and various other researchers received a large grant from the National Science Foundation to create a Center for Biological Timing, with headquarters at the University of Virginia. Emboldened by the financial and intellectual support of the federal government, Takahashi and several colleagues embarked in a very ambitious project to overdo nature and create a circadian mutation of their own in the mouse. And their ambition paid off. After an extensive program of mutagenesis and screening of mutants, the *Clock* gene was identified. *Clock* is a single-locus, semidominant gene that lengthens the circadian period of mice: unaffected mice (+/+) have a circadian period of 23.3 hours, heterozygous mice (+/*Clock*) have a period of 24.4 hours, and homozygotes (*Clock/Clock*) have a period of 27.3 hours.[42] "Mice from Mars!"

Another team of researchers, led by Dr. Gary Pickard from the University of Pennsylvania, conducted a mutagenesis program similar to Takahashi's but was not as fortunate. The mutant gene that they ended up with (*Wheels*, located in chromosome 4) lengthened circadian period in heterozygous mice but was lethal in homozygotes and produced various unrelated phenotypical abnormalities in heterozygotes such as hyperactive bi-directional circling motion, diminished balance, abnormalities in the inner ear, and reduced body size.[43] "Sick Mice from Mars!"

Of course, other researchers of circadian rhythms were not on vacation while Takahashi and Pickard worked on their projects. Three mammalian homologs of the *per* gene were identified by other research teams: *Per1*,[44,45] *Per2*,[46] and *Per3*.[47] RNA levels of these genes were found to exhibit circadian rhythmicity in the SCN and — in the case of *Per1* and *Per2* — to be increased in response to photic stimulation during subjective night (and not during subjective day).[44-48] Not to be outdone, Takahashi's team fully characterized the *Clock* gene (a large transcription unit with 24 exons spanning approximately 100,000 base pairs of DNA)[49,50] and identified another gene in the mouse, *bmal*1, which seems to be essential for circadian rhythmicity.[51] They found that *bmal*1, *Clock*, and *Per*1 are coexpressed in the SCN, and the proteins encoded by the first two (BMAL1 and CLOCK) seem to interact to drive the translation of PER1. Shortly thereafter, another team of researchers found that BMAL1 mRNA exhibits circadian rhythmicity in the SCN of the rat, although its response to light is not phase-dependent in this species.[52] Mammalian homologs of the fruit fly's *tim* gene were also identified and cloned, although mRNA levels of TIM were found not to oscillate daily and not to be affected by photic stimulation.[53-55]

On the other side of the spectrum, an invertebrate homolog of the *Clock* gene was identified in *Drosophila*. As depicted in Figure 12B, CLOCK and CYC (the *Drosophila* homolog of BMAL1) seem to interact as a positive element in a circadian transcriptional loop by stimulating the expression of *per* and *tim*.[56,57] A homolog of the *Clock* gene was also identified in the zebrafish.[58] All of this indicates high structural conservation in the integral components underlying circadian pacemakers in vertebrates and invertebrates, despite obvious interspecies differences. Transcriptional feedback loops composed of negative and positive regulatory elements that alternate in their operation and involve the PER and CLOCK proteins seem to be essential features of circadian rhythmicity in animals. On the other hand, a comparative study of 26 species of the order Lepidoptera (moths and butterflies) revealed that the *per* gene evolves very rapidly, the rate of amino acid substitutions being 5 to 40 times higher than that of two other nuclear genes not involved in circadian rhythmicity.[59] Naturally, detailed comparisons of the molecular mechanisms of circadian rhythmicity in different species cannot be conducted until the mechanisms are fully understood in at least a few species. However, at the pace that research has proceeded in recent years, the occasion may lie in the very near future.

bibliographic notes

1. Crichton, M. (1990). *Jurassic Park*. New York: Knopf.
2. Krauthammer, C. (1997). A special report on cloning. *Time Magazine*, vol. 149, no. 10 (March 10, 1997).
3. Doolittle, R. F., Feng, D. F., Tsang, S., Cho, G., and Little, E. (1996). Determining divergence times of the major kingdoms of living organisms with a protein clock. *Science* 271: 470-477.
4. Starr, K. (1998). *The Starr Report: The Official Report of the Independent Counsel's Investigation of the President*. Rocklin, CA: Prima.
5. Waterston, R. and Sulston, J. E. (1998). The Human Genome Project: Reaching the finish line. *Science* 282: 53-54.
6. Ayer, A.J. (1959). *Logical Positivism*. Chicago: Free Press.
7. Carnap, R. (1966). *Philosophical Foundations of Physics*. New York: Basic Books.

8. Quine, W. V. (1995). *From Stimulus to Science*. Cambridge, MA: Harvard University Press.

9. Crick, F. (1995). *The Astonishing Hypothesis: The Scientific Search for the Soul*. New York: Simon & Schuster.

10. Aronson, B. D., Johnson, K. A., and Dunlap, J. C. (1994). Circadian clock locus *frequency*: Protein enconded by a single open reading frame defines period length and temperature compensation. *Proceedings of the National Academy of Sciences* 91: 7683-7687.

11. Kornhauser, J. M., Nelson, D. E., Mayo, K. E., and Takahashi, J. S. (1990). Photic and circadian regulation of *c-fos* gene expression in the hamster suprachiasmatic nucleus. *Neuron* 5: 127-134.

12. Rusak, B., Robertson, H. A., Wisden, W., and Hunt, S. P. (1990). Light pulses that shift rhythms induce gene expression in the suprachiasmatic nucleus. *Science* 248: 1237-1240.

13. Aronin, N., Sagar, S. M., Sharp, F. R., and Schwartz, W. J. (1990). Light regulates expression of a Fos-related protein in rat suprachiasmatic nuclei. *Proceedings of the National Academy of Sciences* 87: 5959-5962.

14. Ginty, D. D., Kornhauser, J. M., Thompson, M. A., Bading, H., Mayo, K. E., Takahashi, J. S., and Greenberg, M. E. (1993). Regulation of CREB phosphorilation in the suprachiasmatic nucleus by light and a circadian clock. *Science* 260: 238-241.

15. von Gall, C., Duffield, G. E., Hastings, M. H., Kopp, M. D. A., Dehghani, F., Korf, H. W., and Stehle, J. H. (1998). CREB in the mouse SCN: A molecular interface coding the phase-adjusting stimuli light, glutamate, PACAP, and melatonin for clockwork access. *Journal of Neuroscience* 18: 10389-10397.

16. Sutin, E. L. and Kilduff, T. S. (1992). Circadian and light-induced expression of immediate early genes mRNAs in the rat suprachiasmatic nucleus. *Molecular Brain Research* 15: 281-290.

17. Mead, S., Ebling, F. J. P., Maywood, E. S., Humby, T., Herbert, J., and Hastings, M. H. (1992). A nonphotic stimulus causes instantaneous phase advances of the light-entrainable circadian oscillator of the Syrian hamster but does not induce the expression of *c-fos* in the suprachiasmatic nuclei. *Journal of Neuroscience* 12: 2516-2522.

18. Sumová, A., Trávníčková, Z., and Illnerová, H. (1995). Memory on long but not on short days is stored in the rat suprachiasmatic nucleus. *Neuroscience Letters* 200: 191-194.

19. Vuillez, P., Jacob, N., Teclemariam-Mesbah, R., and Pévet, P. (1996). In Syrian and European hamsters, the duration of sensitive phase to light of the suprachaismatic nuclei depends on the photoperiod. *Neuroscience Letters* 208: 37-40.

20. Duffield, G. E., Hastings, M. H., and Ebling, F. J. P. (1998). Investigation into the regulation of the circadian system by dopamine and melatonin in the adult Siberian hamster (*Phodopus sungorus*). *Journal of Neuroendocrinology* 10: 871-884.

21. Zhang, Y., Van Reeth, O., Zee, P. C., Takahashi, J. S., and Turek, F. W. (1993). Fos protein expression in the circadian clock is not associated with phase shifts induced by a nonphotic stimulus, triazolam. *Neuroscience Letters* 164: 203-208.

22. Janik, D. and Mrosovsky, N. (1992). Gene expression in the geniculate induced by a nonphotic circadian phase shifting stimulus. *NeuroReport* 3: 575-578.

23. Colwell, C. S., Kaufman, C. M., and Menaker, M. (1993). Phase-shifting mechanisms in the mammalian circadian system: New light on the carbachol paradox. *Journal of Neuroscience* 13: 1454-1459.

24. Sumová, A., Trávníčková, Z., Mikkelsen, J. D., and Illnerová, H. (1998). Spontaneous rhythm in c-Fos immunoreactivity in the dorsomedial part of the rat suprachiasmatic nucleus. *Brain Research* 801: 254-258.

25. Konopka, R. J. and Benzer, S. (1971). Clock mutants of *Drosophila melanogaster*. *Proceedings of the National Academy of Sciences* 68: 2112-2116.

26. Feldman, J. F. and Hoyle, M. N. (1973). Isolation of circadian clock mutants of *Neurospora crassa*. *Genetics* 75: 605-613.

27. Hall, J. C. (1990). Genetics of circadian rhythms. *Annual Review of Genetics* 24: 659-697.

28. Plautz, J. D., Kaneko, M., Hall, J. C., and Kay, S. A. (1997). Independent photoreceptive circadian clocks throughout *Drosophila*. *Science* 278: 1632-1635.

29. Ewer, J., Frisch, B., Hamblen-Coyle, M. J., Rosbash, M., and Hall, J. C. (1992). Expression of the *period* clock gene within different cell types in the brain of *Drosophila* adults and mosaic analysis of these cell's influence on circadian behavioral rhythms. *Journal of Neuroscience* 12: 3321-3349.

30. Hardin, P. E., Hall, J. C., and Rosbash, M. (1992). Circadian oscillations in *period* gene mRNA levels are transcriptionally regulated. *Proceedings of the National Academy of Sciences* 89: 11711-11715.

31. Edery, I., Rutila, J. E., and Rosbash, M. (1994). Phase shifting of the circadian clock by induction of the Drosophila *period* protein. *Science* 263: 237-240.

32. Scheper, T. O., Klinkenberg, D., Pennartz, C., and van Pelt, J. (1999). A mathematical model for the intracellular circadian rhythm generator. *Journal of Neuroscience* 19: 40-47.

33. Sehgal, A., Price, J. L., Man, B., and Young, M. W. (1994). Loss of circadian behavioral rhythms and *per* RNA oscillations in the *Drosophila* mutant *timeless*. *Science* 263: 1603-1606.

34. Sehgal, A., Rothenfluh-Hilfiker, A., Hunter-Ensor, M., Chen, Y., Myers, M. P., and Young, M. W. (1995). Rhythmic expression of *timeless*: A basis for promoting circadian cycles in *period* gene autoregulation. *Science* 270: 808-810.

35. Gekakis, N., Saez, L., Delahaye-Brown, A. M., Myers, M. P., Sehgal, A., Young, M. W., and Weitz, C. J. (1995). Isolation of *timeless* by PER protein interaction: Defective interaction between *timeless* protein and long-period mutant PER^L. *Science* 270: 811-815.

36. Myers, M. P., Wager-Smith, K., Rothenfluh-Hilfiker, A., and Young, M. W. (1996). Light-induced degradation of TIMELESS and entrainment of the *Drosophila* circadian clock. *Science* 271: 1736-1740.

37. Lee, C., Parikh, V., Itsukaichi, T., Bae, K., and Edery, I. (1996). Resetting the *Drosophila* clock by photic regulation of PER and a PER-TIM complex. *Science* 271: 1740-1744.

38. Aronson, B. D., Johnson, K. A., Loros, J. L., and Dunlap, J. C. (1994). Negative feedback defining a circadian clock: Autoregulation of the clock gene *frequency*. *Science* 263: 1578-1584.

39. Crosthwaite, S. K., Loros, J. J., and Dunlap, J. C. (1995). Light-induced resetting of a circadian clock is mediated by a rapid increase in *frequency* transcript. *Cell* 81: 1003-1012.

40. Crosthwaite, S. K., Dunlap, J. C., and Loros, J. J. (1997). *Neurospora wc-1* and *wc-2*: Transcription, photoresponses, and the origins of circadian rhythmicity. *Science* 276: 763-769.

41. Ralph, M. R. and Menaker, M. (1989). A mutation of the circadian system in golden hamsters. *Science* 241: 1125-1127.

42. Vitaterna, M. H., King, D. P., Chang, A, M., Kornhauser, J. M., Lowrey, P. L., McDonald, J. D., Dove, W. F., Pinto, L. H., Turek, F. W., and Takahashi, J. S. (1994). Mutagenesis and mapping of a mouse gene, *Clock*, essential for circadian behavior. *Science* 264: 719-725.

43. Pickard, G. E., Sollars, P. J., Rinchik, E. M., Nolan, P. M., and Bucan, M. (1995). Mutagenesis and behavioral screening for altered circadian activity identifies the mouse mutant, *Wheels*. *Brain Research* 705: 255-266.

44. Sun, Z. S., Albrecht, U., Zhuchenko, O., Bailey, J., Eichele, G., and Lee, C. C. (1997). RIGUI, a putative mammalian ortholog of the Drosophila *period* gene. *Cell* 90: 1003-1011.

45. Tei, H., Okamura, H., Shineyoshi, Y., Fukuhara, C., Ozawa, R., Hirose, M., and Sakaki, Y. (1997). Circadian oscillation of a mammalian homologue of the Drosophila *period* gene. *Nature* 389: 512-516.

46. Shearman, L. P., Zylka, M. J., Weaver, D. R., Kolakowski, L. F., and Reppert, S. M. (1997). Two *period* homologs: Circadian expression and photic regulation in the suprachiasmatic nuclei. *Neuron* 19: 1261-1269.

47. Takumi, T., Taguchi, K., Miyake, S., Sakakida, Y., Takashima, N., Matsubara, C., Maebayashi, Y., Okumura, K., Takekida, S., Yamamoto, S., Yagita, K., Yan, L., Young, M. W., and Okamura, H. (1998). A light-independent oscillatory gene *mPer3* in mouse SCN and OVLT. *EMBO Journal* 17: 4753-4759.

48. Shigeyoshi, Y., Taguchi, K., Yamamoto, S., Takekida, S., Yan, L., Tei, H., Moriya, T., Shibata, S., Loros, J. J., Dunlap, J. C., and Okamura, H. (1997). Light-induced resetting of a mammalian circadian clock is associated with rapid induction of the *mPer1* transcript. *Cell* 91: 1043-1053.

49. King, D. P., Zhao, Y., Sangoram, A. M., Wilsbacher, L. D., Tanaka, M., Antoch, M. P., Steeves, T. D. L., Vitaterna, M. H., Kornhauser, J. M., Lowrey, P. L., Turek, F. W., and Takahashi, J. S. (1997). Positional cloning of the mouse circadian *Clock* gene. *Cell* 89: 641-653.

50. Antoch, M. P., Song, E. J., Chang, A. M., Vitaterna, M. H., Zhao, Y., Wilsbacher, L. D., Sangoram, A. M., King, D. P., Pinto, L. H., and Takahashi, J. S. (1997). Functional identification of the mouse circadian *Clock* gene by transgenic BAC rescue. *Cell* 89: 655-667.

51. Gekakis, N., Staknis, D., Nguyen, H. B., Davis, F. C., Wilsbacher, L. D., King, D. P., Takahashi, J. S., and Weitz, C. J. (1998). Role of the CLOCK protein in the mammalian circadian mechanism. *Science* 280: 1564-1569.

52. Abe, H., Honma, S., Namihira, M., Tanahashi, Y., Ikeda, M., and Honma, K. (1998). Circadian rhythm and light responsiveness of BMAL1 expression, a partner of mammalian clock gene *Clock*, in the suprachiasmatic nucleus of rats. *Neuroscience Letters* 258: 93-96.

53. Sangoram, A. M., Saez, L., Antoch, M. P., Gekakis, N., Staknis, D., Whiteley, A., Fruechte, E. M., Vitaterna, M. H., Shimomura, K., King, D. P., Young, M. W., Weitz, C. J., and Takahashi, J. S. (1998). Mammalian circadian autoregulatory loop: A *timeless* ortholog and *mPer1* interact and negatively regulate CLOCK-BMAL1-induced transcription. *Neuron* 21: 1101-1113.

54. Zylka, M. J., Shearman, L. P., Levine, J. D., Jin, X. W., Weaver, D. R., and Reppert, S. M. (1998). Molecular analysis of mammalian *timeless*. *Neuron* 21: 1115-1122.

55. Koike, N., Hida, A., Numano, R., Hirose, M., Sakaki, Y., and Tei, H. (1998). Identification of the mammalian homologues of the Drosophila *timeless* gene, *Timeless1*. *FEBS Letters* 441: 427-431.

56. Bae, K., Lee, C., Sidote, D., Chuang, K. Y., and Edery, I. (1998). Circadian regulation of a *Drosophila* homolog of the mammalian *Clock* gene: PER and TIM function as positive regulators. *Molecular and Cellular Biology* 18: 6142-6151.

57. Lee, C. G., Bae, K. H., and Edery, I. (1998). The *Drosophila* CLOCK protein undergoes daily rhythms in abundance, phosphorylation, and interactions with the PER-TIM complex. *Neuron* 21: 857-867.

58. Whitmore, D., Foulkes, N. S., Strahle, U., and Sassone-Corsi, P. (1998). Zebrafish *Clock* rhythmic expression reveals independent peripheral circadian oscillators. *Nature Neuroscience* 1: 701-707.

59. Regier, J. C., Fang, Q. Q., Mitter, C., Peigler, R. S., Friedlander, T. P., and Solis, M. A. (1998). Evolution and phylogenetic utility of the *period* gene in Lepidoptera. *Molecular Biology and Evolution* 15: 1172-1182.

part IV

---

*chronobiology:
business and health care*

chapter eleven

jet lag and shift work

> Following the review of basic processes of circadian
> rhythmicity in the preceding chapters, the application
> of this knowledge to the treatment of jet lag and the
> malaise associated with shift work is discussed here.

In the preceding seven chapters, a lot was said about basic research on
circadian rhythms — about the universality of these rhythms, the characteristics of the clock that controls them, and the neural and molecular mechanisms involved. It is now time to talk about the applications of this vast
knowledge to our daily lives. In this chapter, I will discuss how our knowledge of the biological clock can help us deal with *jet lag* and the malaise
associated with *shift work*.

jet lag

Anyone who has traveled by airplane across several time zones (for instance,
from New York to London) probably has experienced jet lag. Some of the
symptoms that are often reported are fatigue, irritability, inability to concentrate, and difficulty sleeping. The symptoms usually last only a few days (a
little over a week, if many time zones are crossed), but they can be very
distressing, especially for businessmen who must make important business
decisions while mentally impaired by jet lag.

There are three main elements involved in transmeridian flights: the
traveling itself, the crossing of time zones, and the arrival at a different
geographical area. The traveling itself — especially if it is long — may make
you feel bored and tired. The aviation industry is well aware of travelers'
complaints about air transportation (such as airport tumult, barometric pressure changes, immobility, noise, vibration, and radiation) and strives to
improve passenger comfort.[1] However, the stress of an airplane trip is usually nothing that a good night of sleep cannot overcome. Jet lag is a much
more serious condition. After all, if you walk across several time zones
(which is far more exhausting than sitting in an airplane but allows you to

slowly adapt to the successive zones), you do not experience jet lag. Even better, if you fly from New York to Lima in South America (which is a long trip but involves no crossing of time zones), you do not experience jet lag.

Arriving at a different geographical area may also make you feel distressed. A sudden change in temperature, humidity, eating habits, and social customs can certainly make your trip be much less enjoyable than you anticipated. As a matter of fact, depending on one's predisposition, even everyday problems and the minor nuisances of life can be important forms of stress with significant impact on one's health.[2] However, changes in the environment are — for better or for worse — an essential part of traveling. Some people like them, some do not, and this does not seem to have a major effect on jet lag. Indeed, even though the environmental changes can be as radical in a north-south flight as in a west-east flight, jet lag is much stronger in the latter than in the former.

Thus, the major element seems to be the rapid crossing of time zones. In terms of circadian rhythms, what does this mean? Let's consider an eastbound trip across 10 time zones (say, from Los Angeles, California to Athens, Greece). At current commercial air travel speeds, the flight should take about 17 hours. If you leave at 4 pm, you should arrive at 9 am on the next day (according to Los Angeles time, which is also your internal time). However, your 9 am is actually 7 pm in Athens (plus or minus 1 hour, depending on daylight-saving schedules). Particularly if you managed to get some sleep on the plane, you will be quite awake and ready to do business at a time when everyone is going home. Worse, when the Athens clock shows 12 midnight (time to go to sleep), your clock says it is only 2 o'clock in the afternoon. You obviously do not feel sleepy. By the time you do get sleepy (say, midnight in your internal clock, or 10 am in the Athens clock), you are in the middle of a business meeting. No wonder you cannot think clearly!

Naturally, local residents are not jet lagged like you are. They have been in that time zone for quite a while and are fully adapted. So will you, if you can afford to "waste" more than a week sitting by the swimming pool and waiting for your internal clock to be ready to do business in Greece. This is why jet lag is a transient illness — it *will* go away if you do nothing about it. The problem is that usually you cannot wait.

So, the trouble with jet lag is that one is suddenly expected to be awake during one's night and asleep during one's day. Since the person cannot sleep during the day (local night), he or she will be very tired, irritable, and unable to concentrate at night (local day). This sounds very simple, but is it true? Have controlled studies actually been conducted? Yes.

In animal studies, jet lag is simulated by a shift in the light-dark cycle. Advances in the light-dark cycle correspond to eastbound travel (because the sun rises earlier in the East, so that the day starts earlier as one moves eastward instantly), while delays correspond to westbound travel. For instance, in one of my own studies, I recorded the body temperature and activity rhythms of golden hamsters before and after a 6-hour phase advance in the light-dark cycle,[3] which corresponds to a flight from New York to

Paris. Initially, lights were on from 6 am to 8 pm for 10 days. On the 11th day, the light-dark cycle was advanced by 6 hours (lights on from midnight to 2 pm), which was accomplished by shortening the dark cycle on the transition day. The rhythms of body temperature and activity gradually advanced during the following days. Full re-entrainment was attained in 6 days (that is, one day for each hour of the full shift). In a similar study in a different laboratory, it took rats 6 to 7 days to re-entrain after an 8-hour advance in the light-dark cycle.[4]

Studies in which many physiological variables were recorded indicated that different variables may re-entrain at different rates. For instance, in rats subjected to a 12-hour shift of the light-dark cycle, the body temperature rhythm re-entrained in 9 days, while the rhythms of drinking and feeding re-entrained in 7 days.[5] In rhesus monkeys subjected to a 6-hour phase advance, the activity rhythm re-entrained faster than the body temperature rhythm[6] and, in squirrel monkeys subjected to an 8-hour phase delay, the rhythms of drinking, feeding, activity, urination, and body temperature all re-entrained at different rates.[7] As discussed in Chapter 6, differences in phase-shift rates of various rhythms are probably due to differences in the output pathways from the central clock.

A number of studies have been conducted with human subjects in actual and simulated jet lag conditions.[8-12] The time needed for re-entrainment was found to be a little less than 1 day per time zone (say, about 6 days for an 8-hour shift), and differences were found between eastbound and westbound conditions. Of course, small differences in the time needed for re-entrainment may or may not have a significant effect on the traveler's ability to perform tasks at the destination. For instance, it is not clear whether the performance of well-trained athletes is actually impaired by jet lag.[13,14] In an ideal world, the proposal of treatment procedures for jet lag would await a better understanding of the phenomenon. But then, in an ideal world there would be no jet lag. In the real world, there are thousands of travelers who would benefit from a "cure" for jet lag, and it should not be surprising that most research on jet lag has been conducted as part of the search for a cure. So, how can we avoid jet lag?

The simplest way to reduce the effects of jet lag is to be lucky enough to have to do business on the west arm of the trip, rather than on the east arm. Many frequent travelers have found out on their own that jet lag is much less of a problem in westbound flights because they can simply go to sleep a little later (according to their internal clock) on the first night at the destination and will wake up ready for business the following morning. On the other hand, after an eastbound flight, they have to try unsuccessfully to go to bed several hours before their usual bedtime (as discussed above regarding a flight from Los Angeles to Athens). Let us say that you are flying back home from Greece to California. You leave at 8 am, local time (which by now is also your internal time). You fly 17 hours and arrive in Los Angeles on the next day at 1 am, Greece time. Since you crossed 10 time zones, 1 am is actually 3 pm in California (on the previous calendar day). So, it is past your internal bedtime, but you surely can manage to stay awake a few extra hours and go to bed "early," say, at 7 pm (which is 5 am your time). Having gone to bed very late (in terms of your internal clock), you manage to sleep 10 hours and wake up at 5 am (local time). A jog in the park, a long shower, extra commuting time, and you are not too early for the 8:30 am meeting. You may be a little tired at the end of the business day (5 pm local time, 3 am your time), but overall it was not a bad day at all. Of course, if it will take you 6 to 8 days to adapt to the 10-hour change in time zones, you are bound to run into synchronization problems later in the week. However, if you are scheduled to fly back east soon, you may be able to avoid jet lag entirely. Indeed, if you can keep your origination schedule at your destination, you will not experience jet lag at all. The entertainment industry in Las Vegas is very aware of this and keeps the casinos open 24 hours a day, so that travelers from all over the world can gamble at their own origination schedule and can get back home without ever experiencing jet lag.

In some instances, a traveler can avoid jet lag by selecting the best time of the year to travel. When flying from New York to Rio de Janeiro in the winter, the traveler will experience a 3-hour time change, which can be disruptive for business activities. In the summer, however, only a 1-hour change will be experienced — a change that the person is used to experiencing every spring at home. This is due to the fact that, while daylight-saving schedules add one hour to the clocks in each location, summer and winter are in opposite times of the year in the north and south hemispheres (the standard time difference between New York and Rio de Janeiro is 2 hours). If you cannot benefit from this hemispheric discrepancy, you may still be able to reduce jet lag by planning your trip with full awareness of the number of time zones involved (see Exercise No. 10 in the "Doing It Yourself" section).

If you cannot prevent the collision between your internal clock and your destination clock, you may try to convince your hosts to live by your clock (so that they — rather than you — are jet lagged), although it is unlikely that you will succeed. To cause your clock to shift faster than it normally would, you must use one of the various procedures known to phase shift circadian rhythms. None of the procedures studied in the laboratory (such as light pulses, melatonin, and exercise) have been thoroughly tested in real-life jet-lag situations, but many studies have produced encouraging results. Thus, in one study, human volunteers were kept in isolation in the laboratory for two weeks and were subjected to an 8-hour eastbound travel simulation.[15] The body temperature rhythm of subjects who received a 3-hour bright light pulse during subjective morning advanced more than 7 hours during the first 5 days after the shift; the rhythm of subjects who did not receive the light-pulse treatment advanced only 5 hours during the same time. The properly timed light pulse supposedly caused an instantaneous phase-shift of the pacemaker to supplement the gradual phase shift caused by the displacement of the light-dark cycle. Claims of the usefulness of light pulses in the treatment of jet lag have also been made by another research team.[16]

In agreement with these claims, researchers at Cornell University have developed special software to assist people in planning the proper time for the light pulse in preparation for transmeridian flight.[17]

Very popular these days is the use of the hormone melatonin, which — as discussed in Chapter 9 — seems to have the ability to shift the mammalian clock. Since melatonin is not a regulated drug, it can be purchased at natural-food stores and be used by anyone for self medication. The two top-selling popular books on this topic provide plenty of advice on how to obtain and use melatonin[18,19] Articles in the medical literature also support the use of melatonin in the treatment of jet lag, albeit sometimes cautiously.[16,20,21] Other potential treatments include the use of exercise "pulses" to shift the clock after transmeridian flight[22] and the administration of drugs known to have phase-shifting properties,[23] although additional studies are certainly needed.

shift work

I pointed out previously that the central problem in jet lag is that one is suddenly expected to be awake during one's night and asleep during one's day. Well now, except for the fact that there is no travel involved, isn't this the same situation faced by shift workers? Yes, but there is a major difference between the two situations. After a transmeridian flight, the traveler's clock is out of phase with the external clock but it can be re-synchronized in a few days. In shift work, the worker cannot synchronize to the new external clock, as the clock has not actually changed. That is, the worker's work schedule has changed, but the sun still rises and falls at the same time (and so do all other people in the community, except for the few other shift workers). Even worse, shift workers are usually kept on a given shift for only a few days, so that, even if the worker managed to synchronize to the new schedule, he or she would be desynchronized again very soon!

One might argue that rotating shifts were clearly the idea of someone who had no knowledge whatsoever of circadian rhythms. After all, why would someone subject shift workers to the endless torture of repeated phase-shifts just when the workers are re-synchronizing to the new schedule? This criticism would be unfair, however. First, it is unlikely that shift workers actually re-synchronize to the new schedule (remember: the light-dark cycle and the behavior of the rest of the population do not follow the change in schedule of the shift worker). Second, many people will not accept the idea that they have become "night rats." That is, workers themselves demand having days off from the night shift so that, say, they can go to the park with their families on a Sunday afternoon. Thus, not only are shift workers unable to truly adapt to shift work, but they would not want to do it even if they were given the option. Does this mean that shift workers are doomed to have a miserable existence? Not necessarily. Let us have a look at what research in this field has revealed.

Many businesses have three basic types of work shifts: day (8 am to 5 pm), swing (4 pm to 1 am), and graveyard (midnight to 9 am). Workers in the swing and graveyard shifts complain of health problems much more often than workers in the day shift do.[24] As early as 1904, Dr. Francis Benedict conducted a series of careful studies of the body temperature rhythm of shift workers.[25] Being dissatisfied with the use of mercury thermometers in long-term studies, he was one the first physiologists to develop what is today a standard piece of equipment: the thermistor (a temperature-sensitive resistor that allows continuous recording of body temperature based on the variations in electrical resistance). By studying the body temperature rhythm of shift workers and day workers, he found out that even long-term graveyard workers do not synchronize to the night schedule (that is, their body temperature still peaks in the late afternoon, as does that of day workers). A study conducted 94 years later by Drs. Weibel and Brandenberger examined hormonal profiles in nigh shift workers and concurred with Dr. Benedict's conclusion that circadian rhythms do not adapt properly to the reversed work schedule.[26]

Dr. Reinberg's team, in France, looked into the temperature rhythms of shift workers in greater detail. By comparing the temperature rhythms of individual workers with the gravity of their health-related complaints, Reinberg's team found out that the body temperature rhythm of people who are less tolerant of shift work has a smaller amplitude (that is, it has lower peaks and higher troughs).[27] They also found out that rhythms with smaller amplitude are more easily shifted, as assessed on the first night of shift work.[28] Consequently, people who are less tolerant to shift work synchronize to the new schedule more rapidly.[29] This suggests that the malaise associated with shift work may be due to an abrupt shift in the bodily rhythms, and not to the imposition of a shift per se. The conclusion has intuitive appeal if one considers that very slow shifts (say, half an hour per day) are unlikely to result in any distress to the individual. In principle, then, the problem with shift work would not be the fact that one has to work during the night but that one has to make the *transition* from day to night work. Thus, shift work would not be different from jet lag after all. However, Reinberg's team did not follow up the workers' rhythms long enough to ascertain that they actually synchronize to the night schedule. As a matter of fact, even the initial finding of smaller amplitude of the temperature rhythm in people less tolerant to shift work derived from cross-sectional data and may, therefore, result from individual characteristics developed during the course of shift work. Dr. Vidacek's team, in Croatia, conducted a longitudinal study (that is, examining the body temperature rhythms before the individuals ever started shift work and comparing them with health-related complaints after shift work was started) and found out that the body temperature rhythm of people who are less tolerant of shift work has a *larger* amplitude than that of people who are more tolerant[30] (that is, the opposite of Reinberg's finding). Therefore, further studies — especially by different authors — are necessary to elucidate the phenomenon.

As in the case of jet lag, lack of solid scientific knowledge has not prevented attempts to find a cure. Behaviorally, the best way to avoid the malaise associated with shift work is to avoid shift work altogether. Obviously, this is not an option for many workers. For those who must work odd shifts, the best advice is to reduce the number of shift changes and allow the internal clock to adapt to whatever schedule that is required. Of course, one cannot allow the circadian clock to adapt to the new work schedule if the light-dark cycle remains unchanged. A change in the light-dark cycle can be accomplished by exposure to artificial bright light during the night and the use of dark goggles during the day (to avoid light exposure during sunlight hours).[31] As an alternative, one group of researchers has reported that exercise in a stationary cycle ergometer accelerated adaptation to a simulated night shift schedule,[32] although the exercise protocol was rather awkward (15 minutes every hour of the night shift for 3 consecutive days) and the results were not very impressive. Efforts to speed up the shift of the internal clock at the beginning of a new work shift by light pulses[33,34] or melatonin administration[16,35] have had mixed results. Further studies are clearly needed. Such studies will be especially important if humans turn out to be as sensitive to repeated alterations of the daily activity schedule as golden hamsters are. In cardiomyopathic hamsters, animals subjected to a 12-h phase shift of the light-dark cycle on a weekly basis experienced an 11% reduction in life span in reference to animals maintained under a constant light-dark cycle.[36]

bibliographic notes

1. Rayman, R. B. (1997). Passenger safety, health, and comfort: A review. *Aviation Space and Environmental Medicine* 68: 432-440.
2. Kohn, P. M., Lafreniere, K., and Gurevich, M. (1991). Hassles, health, and personality. *Journal of Personality and Social Psychology* 61: 478-482.
3. Refinetti, R. and Menaker, M. (1993). Effects of imipramine on circadian rhythms in the golden hamster. *Pharmacology Biochemistry and Behavior* 45: 27-33.
4. Takamure, M., Murakami, N., Takahashi, K., Kuroda, H., and Etoh, T. (1991). Rapid reentrainment of the circadian clock itself, but not the measurable activity rhythms, to a new light-dark cycle in the rat. *Physiology and Behavior* 50: 443-449.
5. Zerath, E., Holy, X, Lagarde, D., Fernandes, T., Rousselet, D., and Lalouette, A. (1994). Dissociation in body temperature, drinking and feeding rhythms following a light-dark cycle inversion in the rat. *Medical Science Research* 22: 53-55.
6. Tapp, W. N. and Natelson, B. H. (1989). Circadian rhythms and patterns of performance before and after simulated jet lag. *American Journal of Physiology* 257: R796-R803.
7. Moore-Ede, M. C., Kass, D. A., and Herd, J. A. (1977). Transient circadian internal desynchronization after light-dark phase shift in monkeys. *American Journal of Physiology* 232: R31-R37.

8. Elliott, A. L., Mills, J. N., Minors, D. S., and Waterhouse, J. M. (1972). The effect of real and simulated time-zone shifts upon the circadian rhythms of body temperature, plasma 11-hydroxycorticosteroids, and renal excretion in human subjects. *Journal of Physiology* 221: 227-257.

9. Wever, R. A. (1980). Phase shifts of human circadian rhythms due to shifts of artificial zeitgebers. *Chronobiologia* 7: 303-327.

10. Folkard, S., Minors, D. S., and Waterhouse, J. M. (1991). "Demasking" the temperature rhythm after simulated time zone transitions. *Journal of Biological Rhythms* 6: 81-91.

11. Monk, T. H., Moline, M. L., and Graeber, R. C. (1988). Inducing jet lag in the laboratory: Patterns of adjustment to an acute shift in routine. *Aviation Space and Environmental Medicine* 59: 703-710.

12. Binkley, S. (1994). Wrist motion rhythm phase shifts in travelers may differ from changes in time zones. *Physiology and Behavior* 55: 967-970.

13. O'Connor, P. J., Morgan, W. P., Koltyn, K. F., Raglin, J. S., Turner, J. G., and Kalin, N. H. (1991). Air travel across four time zones in college swimmers. *Journal of Applied Physiology* 70: 756-763.

14. Reilly, T., Atkinson, G., and Waterhouse, J. (1997). Travel fatigue and jet-lag. *Journal of Sports Science* 15: 365-369.

15. Honma, K. I., Honma, S., Nakamura, K., Sasaki, M., Endo, T., and Takahashi, T. (1995). Differential effects of bright light and social cues on reentrainment of human circadian rhythms. *American Journal of Physiology* 268: R528-R535.

16. Lewy, A. J., Ahmed, S., and Sack, R. L. (1996). Phase shifting the human circadian clock using melatonin. *Behavioural Brain Research* 73: 131-134.

17. Houpt, T. A., Boulos, Z., and Moore-Ede, M. C. (1996). MidnightSun: Software for determining light exposure and phase-shifting schedules during global travel. *Physiology and Behavior* 59: 561-568.

18. Reiter, R. J. and Robinson, J. (1996). *Melatonin: Breakthrough Discoveries that Can Help You.* New York: Bantam.

19. Pierpaoli, W., Davies, O., and Regelson, W. (1995). *The Melatonin Miracle: Nature's Age-Reversing, Disease-Fighting, Sex-Enhancing Hormone.* New York: Simon & Schuster.

20. Comperatore, C. A., Lieberman, H. R., Kirby, A. W., Adams, B., and Crowley, J. S. (1996). Melatonin efficacy in aviation missions requiring rapid deployment and night operations. *Aviation Space and Environmental Medicine* 67: 520-524.

21. Arendt, J., Skene, D. J., Middleton, B., Lockley, S. W., and Deacon, S. (1997). Efficacy of melatonin treatment in jet lag, shift work, and blindness. *Journal of Biological Rhythms* 12: 604-617.

22. Shiota, M., Sudou, M., and Ohshima, M. (1996). Using outdoor exercise to decrease jet lag in airline crewmembers. *Aviation Space and Environmental Medicine* 67: 1155-1160.

23. Van Reeth, O. and Turek, F. W. (1987). Adaptation of circadian rhythmicity to shift in light–dark cycle accelerated by a benzodiazepine. *American Journal of Physiology* 253: R204-R207.

24. Singer, B., Terborg, J., and Mayer, S. (1994). Attitudinal, circadian, circumstantial, and subject selection explanations of shiftwork effects on health. *Journal of Occupational Medicine* 36: 66-69.

25. Benedict, F. G. (1904). Studies in body temperature. I. Influence of the inversion of the daily routine. *American Journal of Physiology* 11: 145-169.

26. Weibel, L. and Brandenberger, G. (1998). Disturbances in hormonal profiles of night workers during their usual sleep and work times. *Journal of Biological Rhythms* 13: 202-208.
27. Andlauer, P., Reinberg, A., Fourré, L., Battle, W., and Duverneuil, G. (1979). Amplitude of the oral temperature circadian rhythm and the tolerance to shift-work. *Journal de Physiologie* 75: 507-512.
28. Reinberg, A., Vieux, N., Ghata, J., Chaumont, A. J., and Laporte, A. (1978). Is the rhythm amplitude related to the ability to phase-shift circadian rhythms of shift-workers? *Journal de Physiologie* 74: 405-409.
29. Reinberg, A., Andlauer, P., Guillet, P., and Nicolai, A. (1980). Oral temperature, circadian rhythm amplitude, ageing and tolerance to shift-work. *Ergonomics* 23: 55-64.
30. Vidacek, S., Radosevic-Vidacek, B., Kaliterna, L., and Prizmic, Z. (1993). Individual differences in circadian rhythm parameters and short-term tolerance to shiftwork: A follow-up study. *Ergonomics* 36: 117-123.
31. Eastman, C. I., Stewart, K. T., Mahoney, M. P., Liu, L., and Fogg, L. F. (1994). Dark goggles and bright light improve circadian rhythm adaptation to night-shift work. *Sleep* 17: 535-543.
32. Eastman, C. I., Hoese, E. K., Youngstedt, S. D., and Liu, L. (1995). Phase-shifting human circadian rhythms with exercise during the night shift. *Physiology and Behavior* 58: 1287-1291.
33. Gallo, L. C. and Eastman, C. I. (1993). Circadian rhythms during gradually delaying and advancing sleep and light schedules. *Physiology and Behavior* 53: 119-126.
34. Eastman, C. I., Liu, L., and Fogg, L. F. (1995). Circadian rhythm adaptation to simulated night shift work: Effect of nocturnal bright-light duration. *Sleep* 18: 399-407.
35. James, M., Tremea, M. O., Jones, J. S., and Krohmer, J. R. (1998). Can melatonin improve adaptation to night shift? *American Journal of Emergency Medicine* 16: 367-370.
36. Penev, P. D., Kolker, D. E., Zee, P. C., and Turek, F. W. (1998). Chronic circadian desynchronization decreases the survival of animals with cardiomyopathic heart disease. *American Journal of Physiology* 275: H2334-H2337.

chapter twelve

sleep disorders

The influence of the circadian system on human sleep
and its implications for the treatment of sleep disorders
are examined.

We saw in Chapter 4 that the existence of a daily rhythm of activity is
pervasive in the animal kingdom. In many animals, including humans, the
rest phase of the activity rhythm includes the phenomenon of sleep. The
amount of time spent sleeping each day varies greatly among the vertebrates:
about 3 hours in horses, 8 hours in humans, and 15 hours in cats.[1] A study
in which four people recorded their bedtimes and waketimes every day for
a whole year showed a persistent 24-hour periodicity (not surprisingly, considering the presence of multiple zeitgebers) with small alterations in phase
in response to the beginning and end of daylight savings and to travel
overseas.[2] A survey of college students conducted in California in 1916
indicated that average bedtime and waketime were 11 pm and 7 am, respectively,[3] while a survey conducted in Virginia in 1995 indicated that average
bedtime and waketime were 1:30 am and 9 am.[4] Thus, although college
students are often said to be chronically sleep-deprived, most of them actually sleep an average of 7 to 8 hours per night. As expected, however,
waketimes are strongly determined by class schedules, so that a weekly
pattern of waketimes (with delayed waketimes on weekends) can be
observed in the records of many students.[4,5]

Considering our everyday experience of wanting to sleep when we are
tired and of feeling rested after a good night of sleep, it is rather intuitive
that sleep has a restorative function and that sleep deprivation causes irritability, increased anxiety, and impaired concentration.[6] As determined in
rats, prolonged sleep deprivation can be fatal.[7] To say that sleep serves a
restorative function, however, implies that something has to be restored. If
sleep allows the dissipation of some somnogenic (sleep-inducing) substance
that was accumulated during wake hours, then what is this substance? There
is not yet a definitive answer to this question, but a good candidate is
adenosine. Adenosine, as you may recall, is the A in the acronym ATP (ade-

nosine tri-phosphate). ATP is the energetic currency of the body — the molecule that transports the energy derived from the breakdown of nutrients to individual cells. Adenosine is a metabolite of ATP, which means that its accumulation is proportional to the amount of energy expended by the body. The more energy is expended (that is, the more active the organism is), the greater is the accumulation of adenosine. Therefore, adenosine meets a basic requirement of a somnogenic substance. Research in cats has shown that adenosine concentration in the brain increases during wakefulness and that infusion of adenosine into the brain induces sleep.[8,9] Further corroborative evidence should be available soon.

So, we feel sleepy as a consequence of the accumulation of a somnogenic substance, which may be adenosine. But what about circadian rhythmicity? It was mentioned briefly in Chapter 4 that sleepiness is strongly associated with the trough of the body temperature rhythm, which means that people feel like sleeping (and often do sleep) around the particular time of the circadian day when the body temperature rhythm is at its low point.[10–14] That is, sleepiness is dependent on circadian time, not on sleep deprivation. You may have experienced this yourself if you ever had to stay up all night. In the early evening, you felt reasonably energetic. As the night progressed past your bedtime, you became progressively sleepier. However, by early morning you started to feel awake again, even though you did not sleep all night. At lunch time, you felt a little sleepy again, but the situation improved in the afternoon. Finally, when the night came, you felt truly tired and slept like a log.

Now, what can we do with these conflicting pieces of evidence? Is sleep controlled by a restorative process or by a circadian process? Actually, by both. Sleep is a restorative process *gated* by the circadian system.[15] The longer you stay awake, the greater is your potential to sleep; however, the potential is converted into actual sleep propensity only when the circadian gate is open. Conversely, if the gate is open (if it is that time of the circadian cycle when your body temperature is falling) but you have just woken up, there will be little sleep propensity.

To avoid a common misunderstanding, I should emphasize that the fall in body temperature with which sleep is associated[16,17] is only a marker of the phase of the circadian pacemaker. Low body temperature per se is not sleep inducing. Large reductions in ambient temperature (which can cause a fall in body temperature) do not induce sleep; on the contrary, they impair sleep.[18] As a matter of fact, if there is a causal link between body temperature and sleep, it is in the opposite direction: falling asleep causes a small decrease in body temperature superimposed on the larger decrease determined by the circadian system.[19–21] Also, during the deeper stages of sleep, the homeostatic control of body temperature is relaxed,[22–24] which can lead to a further fall in body temperature if ambient temperature is low. It has been reported that pulses of bright light presented at bedtime reduce the subsequent fall in body temperature,[25–27] although the significance of this finding is not clear.

While most of us can get a good night of sleep if our work schedule allows, many people face chronic sleep problems. Sleep disorders include insomnia (inability to sleep during the usual sleep time), narcolepsy (intrusion of sleep episodes into wake time), somnambulism (sleepwalking), and night terrors. Some instances of insomnia, such as those caused by sleep apnea, have no obvious relationship with circadian rhythms. Other instances, however, could potentially be due to disorders of the circadian system; among them are the so-called Delayed Sleep Phase Syndrome (DSPS) and Advanced Sleep Phase Syndrome (ASPS).

ASPS and DSPS seem to be simple cases of an abnormal phase angle of entrainment. ASPS patients feel very sleepy abnormally early in the evening and then wake up very early in the morning (say, 2 or 3 am). Conversely, DSPS patients have difficulty falling asleep before 2 or 3 am and, consequently, will not wake up spontaneously until late in the morning. DSPS is more common than ASPS, afflicting between 1 and 4 people out of 1000.[28] Neither case involves actual lack of sleep if the patients can afford to set their work schedules according to their sleep habits, and studies of DSPS patients using electroencephalographic recording revealed no abnormalities in sleep structure.[29,30] However, most people do not have the luxury of determining their own work schedule. Consequently, DSPS and ASPS regularly result in chronic lack of sleep, which is experienced as sleep-onset insomnia in DSPS patients and as sleep-termination insomnia in ASPS patients.

If DSPS and ASPS actually derive from disturbances in the phase angle of entrainment (rather than from some other abnormality that only secondarily affects the phase angle), then our knowledge of circadian rhythms should allow us to understand the etiology of the disease. We saw in Chapters 5 and 6 that entrainment depends on three variables: the period of the pacemaker (τ), the period of the zeitgeber (T), and the PRC of the pacemaker. The period of the zeitgeber is presumably 24 hours for all patients living in the real world, so that we need only worry about the other two variables. If t is shorter than 24 hours, the activity rhythm will be phase advanced in reference to the zeitgeber (that is, the person will want to wake up earlier); therefore ASPS could result from τ being about 20 hours. Conversely, if τ is longer than 24 hours, the activity rhythm will be phase delayed; this is the case for most people (since τ is usually 25 hours) but could be exaggerated in DSPS patients (say, τ = 28 hours). Just like the activity of the nocturnal hamster with a short τ was displaced into the daylight (Figure 6 in Chapter 6), the activity of the diurnal human with a long τ would will be displaced into the nighttime. Alternatively (or in combination with an abnormal τ), the PRC could be affected. Thus, the relative phase delay of DSPS patients could be due to a reduced amplitude of the phase-advance region of the PRC. A reduced amplitude would require that a wider section of the phase-advance region be photically stimulated, which could only be accomplished by starting the stimulation earlier during subjective night and terminating earlier during subjective day, thus extending the duration between lights off and bedtime. The combination of a small reduction in the amplitude of the PRC and a small lengthening of τ would produce equivalent results.

Of course, this is all hypothetical. It depends on whether DSPS and ASPS actually derive from disturbances in the phase angle of entrainment. Do they? The bedtime and waketime data suggest that they do. However, we saw previously that the wake-sleep rhythm in humans can be very labile and not reflect the state of the circadian pacemaker. Have other rhythms, such as those of body temperature and melatonin secretion, been recorded in patients with sleep phase disorders? Yes, although only in a couple of studies.[31,32] Have the circadian period and PRC amplitude been determined in patients with sleep phase disorders? Not really.[33] Despite this scarcity of data about the etiology of sleep phase disorders, several forms of treatment have been employed. These have included chronotherapy, treatment with bright light exposure, and treatment with melatonin administration.

All three of these forms of treatment have the peculiarity of not involving controlled drugs and, therefore, not requiring a prescribing physician. In particular, chronotherapy is closer to a businesslike time-management initiative than to a medical treatment. It involves the realization that the problem at hand is simply one of doing the right thing at the right time. If the person is going to bed too late (and losing sleep only because he or she must wake up early in the morning to go to work), then the solution consists in moving the bedtime to an earlier time. Of course, the person cannot come up with a solution by him/herself — otherwise, the problem would not exist to start with. However, considering that the usual human free-running period is longer than 24 hours, the adjustment in bedtime can be made if its direction is reversed: rather than trying to go to bed earlier, the person should delay the bedtime a few hours each day until he or she reaches the desired bedtime.[34] The new phase angle of entrainment is unlikely to persist if the old phase angle was determined by abnormalities in τ or PRC, but it may very well persist if the true etiology is different. Except for the requirement of a week off from work to implement the transition, the procedure is very little disruptive and can be performed by the patient him/herself without the consultation of a physician (or of a business advisor, or any third party at all). Therefore, it provides the convenience of a self-help method.

Bright light treatment is based on the concept of light-induced phase shifting of the circadian pacemaker (as discussed in Chapters 5 and 6), even though it is still not clear whether the therapeutic action of bright light is always of a non-parametric nature (see Chapter 13). Thus, if a patient is entrained to the 24-hour world with a chronic phase delay (DSPS), properly-timed light pulses presented during the phase-advance region of the PRC should cause a corrective phase advance. Successful treatment of DSPS patients with phototherapy has been reported.[35,36] Of course, this is another case where the long-term success of a procedure contradicts the rationale on which it is based. If the circadian pacemaker was abnormally entrained because of its abnormal properties, then the phase shift caused by phototherapy should be very short-lived and, consequently, should be of very little clinical use. In other words, if the rationale for phototherapy is correct, then patients should require daily light pulses throughout their lives. The same is true about the apparent success in the use of melatonin administration (which is currently classified as a food supplement, not as a prescription drug) in the treatment of sleep phase disorders.[37,38] It is possible that the therapeutic action of melatonin is dependent not on a non-parametric effect on the circadian pacemaker but rather on a parametric hypnotic (sleep-inducing) effect, although not even this is clear.[39]

Needless to say, the fact that physicians employ some form of treatment (and seem to obtain positive results) does not imply that the treatment is effective. As will be elaborated further in the following Chapter, physicians have the obligation to try to help their patients the best they can. They cannot afford to wait until all scientific doubts are eliminated. To complicate things further, it seems that even clinical experts are often ignorant of significant recent developments in clinical research.[40]

The extent to which placebo effects are responsible for the improvements attained by the treatment of sleep phase disorders is unknown. From a scientific perspective, the circadian etiology of ASPS and DSPS is largely hypothetical, and the efficacy of procedures used to treat these disorders may or may not exceed that of placebo. From a clinical perspective, however, potential solutions to the patients' predicaments are being offered — and oftentimes this is as much as medicine can hope for.

bibliographic notes

1. Campbell, S. S. and Tobler, I. (1984). Animal sleep: A review of sleep duration across phylogeny. *Neuroscience and Biobehavioral Reviews* 8: 269-300.
2. Binkley, S., Begona, M., Crawford, D., and Mosher, K. (1990). Human daily rhythms measured for one year. *Physiology and Behavior* 48: 293-298.
3. Gates, A. I. (1916). Diurnal variations in memory and association. *University of California Publications in Psychology* 1: 323-344.
4. Refinetti, R. (1995). Persistence of synchronization of the daily rhythms of body temperature and sleep-wake in college students. *Biological Rhythm Research* 26: 532-540.
5. Binkley, S., Tome, M. B., and Mosher, K. (1989). Weekly phase shifts of rhythms self-reported by almost feral human students in the USA and Spain. *Physiology and Behavior* 46: 423-427.
6. Dement, W. (1960). The effect of dream deprivation. *Science* 131: 1705-1707.
7. Everson, C. A., Bergmann, B. M., and Rechtschaffen, A. (1989). Sleep deprivation in the rat. II. Total sleep deprivation. *Sleep* 12: 13-21.
8. Porkka-Heiskanen, T., Strecker, R. E., Thakkar, M., Bjørkum, A. A., Greene, R. W., and McCarley, R. W. (1997). Adenosine: A mediator of the sleep-inducing effects of prolonged wakefulness. *Science* 276: 1265-1268.
9. Portas, C. M., Thakkar, M., Rainnie, D. G., Greene, R. W., and McCarley, R. W. (1997). Role of adenosine in behavioral state modulation: A microdialysis study in the freely moving cat. *Neuroscience* 79: 225-235.

10. Zulley, J., Wever, R., and Aschoff, J. (1981). The dependence of onset and duration of sleep on the circadian rhythm of rectal temperature. *Pflügers Archiv* 391: 314-318.

11. Czeisler, C. A., Weitzman, E. D., Moore-Ede, M. C., Zimmerman, J. C., and Knauer, R. S. (1980). Human sleep: Its duration and organization depend on its circadian phase. *Science* 210: 1264-1267.

12. Zulley, J. and Campbell, S. S. (1985). Napping behavior during "spontaneous internal desynchronization": Sleep remains in synchrony with body temperature. *Human Neurobiology* 4: 123-126.

13. Gillberg, M. and Åkerstedt, T. (1982). Body temperature and sleep at different times of the day. *Sleep* 5: 378-388.

14. Dijk, D. J. and Czeisler, C. A. (1994). Paradoxical timing of the circadian rhythm of sleep propensity serves to consolidate sleep and wakefulness in humans. *Neuroscience Letters* 166: 63-68.

15. Daan, S., Beersma, D. G. M., and Borbély, A. A. (1984). Timing of human sleep: Recovery process gated by a circadian pacemaker. *American Journal of Physiology* 246: R161-R178.

16. Campbell, S. S. and Zulley, J. (1988). Napping as a biological rhythm: Disentrainment of the human sleep-wake system. In: Koella, W. P., Obál, F., Schulz, H., and Visser, P. (Eds.). *Sleep '86*. New York: Gustav Fischer.

17. Campbell, S. S. and Broughton, R. J. (1994). Rapid decline in body temperature before sleep: Fluffing the physiological pillow? *Chronobiology International* 11: 126-131.

18. Muzet, A., Libert, J. P., and Candas, V. (1984). Ambient temperature and human sleep. *Experientia* 40: 425-429.

19. Obál Jr., F., Rubicsek, G., Alföldi, P., Sári, G., and Obál, F. (1985). Changes in the brain and core temperatures in relation to the various arousal states in rats in the light and dark periods of the day. *Pflügers Archiv* 404: 73-79.

20. Barrett, J., Lack, L., and Morris, M. (1993). The sleep-evoked decrease of body temperature. *Sleep* 16: 93-99.

21. Almirall, H., Aguirre, A., Rial, R. V., Daurat, A., Foret, J., and Benoit, O. (1993). Temperature drop and sleep: Testing the contribution of SWS in keeping cool. *NeuroReport* 5: 177-180.

22. Shapiro, C. M., Moore, A. T., Mitchell, D., and Yodaiken, M. L. (1974). How well does man thermoregulate during sleep? *Experientia* 30: 1279-1281.

23. Sagot, J. C., Amoros, C., Candas, V., and Libert, J. P. (1987). Sweating responses and body temperatures during nocturnal sleep in humans. *American Journal of Physiology* 252: R462-R470.

24. Dewasmes, G., Bothorel, B., Candas, V., and Libert, J. P. (1997). A short-term poikilothermic period occurs just after paradoxical sleep onset in humans: Characterization of changes in sweating effector activity. *Journal of Sleep Research* 6: 252-258.

25. Dijk, D. J., Cajochen, C., and Borbély, A. A. (1991). Effect of a single 3-hour exposure to bright light on core body temperature and sleep in humans. *Neuroscience Letters* 121: 59-62.

26. Badia, P., Myers, B., Boecker, M., Culpepper, J., and Harsh, J. R. (1991). Bright light effects on body temperature, alertness, EEG and behavior. *Physiology and Behavior* 50: 583-588.

27. Myers, B. L. and Badia, P. (1993). Immediate effects of different light intensities on body temperature and alertness. *Physiology and Behavior* 54: 199-202.

28. Vignau, J., Dahlitz, M., Arendt, J., English, J., and Parkes, J. D. (1993). Biological rhythms and sleep disorders in man: The delayed sleep phase syndrome. In: Wetterberg, L. (Ed.). *Light and Biological Rhythms in Man.* New York: Pergamon, pp. 261-271.

29. Weitzman, E. D., Czeisler, C. A., Coleman, R. M., Spielman, A. J., Zimmerman, J. C., and Dement, W. (1981). Delayed sleep phase syndrome: A chronobiological disorder with sleep-onset insomnia. *Archives of General Psychiatry* 38: 737-746.

30. Alvarez, B., Dahlitz, M. J., Vignau, J., and Parkes, J. D. (1992). The delayed sleep phase syndrome: Clinical and investigative findings in 14 subjects. *Journal of Neurology Neurosurgery and Psychiatry* 55: 665-670.

31. Lack, L. and Wright, H. (1993). The effect of evening bright light in delaying the circadian rhythms and lengthening the sleep of early morning awakening insomniacs. *Sleep* 16: 436-443.

32. Okawa, M., Uchiyama, M., Shirakawa, S., Takahashi, K., Mishima, K., and Hishikawa, Y. (1993). Favorable effects of combined treatment with viatmin B_{12} treatment and bright light for sleep-wake rhythm disorders. In: Kumar, V. M., Malick, H. N., and Nayar, U. (Eds.). *Sleep-Wakefulness.* New Delhi: Wiley Eastern, pp. 71-77.

33. Terman, M., Lewy, A. J., Dijk, D. J., Boulos, Z., Eastman, C, I., and Campbell, S. S. (1995). Light treatment for sleep disorders: Consensus report. IV. Sleep phase and duration disturbances. *Journal of Biological Rhythms* 10: 135-147.

34. Czeisler, C. A., Richardson, G., Coleman, R., Zimmerman, J., Moore-Ede, M., Dement, W., and Weitzman, E. (1981). Chronotherapy: Resetting the circadian clocks of patients with delayed sleep phase insomnia. *Sleep* 4: 1-21.

35. Lewy, A. J., Sack, R. L., and Singer, C. M. (1985). Treating phase typed chronobiologic sleep and mood disorders using appropriately timed bright artificial light. *Psychopharmacological Bulletin* 21: 368-372.

36. Rosenthal, N. E., Joseph-Vanderpool, J. R., Levendosky, A. A., Johnston, S. H., Allen, R., Kelly, K. A., Souêtre, E., Schultz, P. M., and Starz, K. (1990). Phase-shifting effects of bright morning light as treatment for delayed sleep phase syndrome. *Sleep* 13: 354-361.

37. Dahlitz, M., Alvarez, B., Vignau, J., English, J., Arendt, J., and Parkes, J. D. (1991). Delayed sleep phase syndrome response to melatonin. *Lancet* 337: 1121-1123.

38. Tzischinsky, O., Dagan, Y., and Lavie, P. (1993). The effects of melatonin on the timing of sleep in patients with delayed sleep phase syndrome. In: Touitou, Y., Arendt, J., and Pévet, P. (Eds.). *Melatonin and the Pineal Gland.* New York: Elsevier, pp. 351-354.

39. Dawson, D. and Encel, N. (1993). Melatonin and sleep in humans. *Journal of Pineal Research* 15: 1-12.

40. Antman, E. M., Lau, J., Kupelnick, B., Mosteller, F., and Chalmers, T. C. (1992). A comparison of results of meta-analyses of randomized control trials and recommendations of clinical experts: Treatments for myocardial infarction. *Journal of the American Medical Association* 268: 240-248.

chapter thirteen

depression and seasonal affective disorder

> Affective disorders (such as depression and seasonal affective disorder) seem to be related to a dysfunction of the circadian system, which implies that they may be treatable by behavioral means as an alternative to antidepressant drugs.

The *PDR Medical Dictionary* defines medicine as "the art of preventing or curing disease."[1] Although scientific knowledge can greatly help the prevention and cure of disease, the art of preventing and curing must be more than the application of scientific knowledge to "real-world" situations. The central mission of medicine is certainly not to advance basic scientific knowledge but to improve the patient's health. Even when no scientific knowledge about a given condition is available, the physician is expected to do his or her best to alleviate the patient's ailment. In this sense, even a placebo effect is a meaningful effect. It could be reasonably argued that placebo medication should be less expensive than medication whose pharmacological action is known, but there is no reason to discard the clinical relevance of placebo medication.[2] This is especially true when psychiatric conditions such as affective disorders (depression, mania, and bipolar disorders) are treated through psychotherapy. Indeed, in the strict sense of the term, all psychotherapy is placebo medication, as the pharmacological action of psychotherapy is still unknown. As a matter of fact, even the behavioral action of psychotherapy is not well understood. Evaluation of the efficacy of psychotherapy is complicated by the fact that traditionally only a small fraction of people exhibiting deviant behaviors go through psychotherapy.[3-4] Thus, most cases of "anti-social personality" (i.e., psychotics, criminals, juvenile delinquents, etc.) are isolated in special institutions and do not undergo psychotherapy (i.e., they receive other types of treatment or no treatment at all). Thirty to sixty percent of those who consult a psychotherapist and are accepted for treatment stop coming back before the therapist thinks they should do so. Of those who stay, about two thirds are considered to show improvement. Thus, out of 100 people with deviant behaviors, only 40 undergo psychotherapy, 20 remain under treatment long enough, and 14 show improvement. Most of us would not call a 14/100 ratio a success. But we would not call it a failure either. After all, it is not the therapist's fault that patients are not sent in or do not wish to stay. What about the 2/3 improvement ratio? Well, this ratio may not be a success either. Since many people who are institutionalized do not undergo therapy, there is a natural control group against which to evaluate the efficacy of psychotherapeutic treatment. In 1952, H. J. Eysenck compared a group of adult neurotic patients undergoing psychotherapy with a similar group of patients not exposed to psychotherapy. His conclusion was straightforward: two thirds of a group of adult neurotic patients will show some improvement within two years after the beginning of the illness irrespective of whether they underwent psychotherapy or not.[5] A few years later, a similar study on children reached a similar conclusion.[6] Maybe not surprisingly, the community of psychotherapists initially did not pay much attention to these studies.[7] By the 1980s, however, it was figured out that psychotherapy does help some patients, although the specifics of the therapy are almost irrelevant.[8-9] That is, in contrast to the therapist's belief (and the belief of most patients), the therapeutic value of psychotherapy resides in simply talking about one's problems to someone else — and not in learning how to "handle repressed feelings," or how to "establish effective contingencies of reinforcement", or how to "manage motivational needs." Thus, the therapist's role is not that of a "scientific practitioner" but that of a "personal consultant" who advises a client the same way that a business consultant advises a business client. This "demystification" of psychotherapy was at the heart of my own consulting practice.[10] It is consistent with the idea of "social construction" of the concept of mental illness.[11,12]

Since the welfare of the patient is the central goal of medical practice, it is not surprising that knowledge about circadian rhythms has been used to develop therapeutic procedures in advance of full scientific understanding of the processes involved. Several authors have suggested that disruption in circadian organization is an etiological (causal) factor in mental illness — particularly affective disorders.[13-17] Although affective disorders usually do not involve the breakdown in logical thinking that is characteristic of other forms of mental illness popularly associated with the concept of "madness," they can nevertheless be very incapacitating. Severe depression can easily lead to suicide. Of course, everyone is subject to occasional alterations in mood elicited by both internal and external factors (such as losing one's job, going through a divorce, etc.), and usually there is nothing pathological in it. Although overall happiness seems to be related much more to one's constitution than to events in one's life,[18] small mood alterations are very common. As a matter of fact, mood seems to vary more or less regularly along the year, as shown in Figure 13, which is based on the results of a

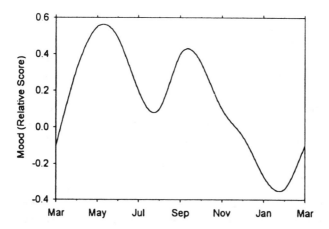

Figure 13.   Annual variation in mood of healthy volunteers in New York City. Notice a small decrease in mood during mid-summer and a greater decrease in the winter.

survey conducted in New York City.[19] However, alterations in mood are sometimes strong enough to be diagnosed as depression or mania and, therefore, require medical attention.

Clinically depressed patients exhibit various abnormalities in circadian rhythms,[20-25] although the question always remains whether the circadian abnormalities are the cause or merely symptoms of the disease. In order to determine whether the correlation between circadian disruption and depression implies causation, experimental research in laboratory animals has been conducted. "Animal models" of depression have several limitations;[26] however, if one assumes that depression is the result of circadian disorganization, and it is known that antidepressant drugs relieve depression, then one must expect that antidepressants will affect circadian organization. Although imipramine (Tofranil) — the first widely used antidepressant — was shown not to affect circadian rhythms in laboratory animals,[27-29] other drugs such as lithium (Eskalith, Lithobid),[30-33] fluoxetine (Prozac),[34] clorgyline (an experimental antidepressant),[35] and rubidium (also an experimental antidepressant)[36] have been found to have small but significant effects. Thus, the hypothesis that circadian disruption causes depression remains unrefuted. It can hardly be said, however, that causality has been demonstrated. Real proof would consist of inducing depression in human subjects by experimental disruption of circadian rhythmicity and relieving the symptoms by restoring circadian organization. Of course, such an experiment would be unethical, if not criminal.

A major roadblock in the understanding of the etiology of depression (and other modalities of mental illness) is that, despite great progress in recent years, we know very little about how the brain processes mental activity. At some point, it seemed that genes responsible for manic depression and alcoholism had been identified, but it all turned out to be a false alarm.[37,38] Researchers have identified brain areas that seem to be altered in schizophrenic patients, but they are still far from identifying the neuroanatomical correlates of the disease.[39,40]

Since the antidepressant imipramine increases activity at synapses that utilize norepinephrine and dopamine as neurotransmitters, it was initially conjectured that depression might be the result of a deficiency in norepinephrine and dopamine.[41] With the introduction of Prozac — which inhibits the synaptic re-uptake of serotonin, thus increasing activity in serotonergic synapses — it was conjectured that depression might be the result of a deficiency in serotonin. More recently, Merck Laboratories have developed a new antidepressant (MK-869) that blocks synaptic receptors for the neurotransmitter substance P and does not seem to affect receptors for serotonin or norepinephrine.[42] Thus, we can produce drugs to treat depression — and they do seem to work well — but we can hardly say that we know how they really work. In this context, it is not surprising that electroconvulsive therapy (the medical equivalent of an electrical technician's kick on the TV set) has been rehabilitated in the treatment of affective disorders.[43]

A particular type of affective disorder, Seasonal Affective Disorder (SAD), deserves special attention because of its rhythmic nature. SAD is a cyclic illness characterized by recurrent episodes of fall and winter depression alternating with periods of remission during the spring and summer.[44,45] Because of this seasonal rhythmicity, it was suggested early on that SAD might be under the influence of daylength, as many annual rhythms in animals are (see Chapter 3). Indeed, although weather conditions were found to affect patients' self-rating of energy level, this effect was smaller than those of seasonal changes.[46] Accordingly, attempts were made to treat SAD by artificially extending the duration of daylight during the winter, typically by providing bright light stimulation for several hours in the early morning and early evening. Various authors reported mood improvements in SAD patients subjected to this form of treatment.[47-51] However, it soon became clear that the effects of light therapy were not due to the extension of daylength.[52-57] For instance, it was found that a single light stimulation in the morning was just as effective as the morning-evening combination and, as a matter of fact, that a single stimulation in the early afternoon (which, of course, does not extend the daylength) was also effective.[55] Attention was then focused on the potential phase-shifting effects of phototherapy. As mentioned in Chapter 6, appropriately timed light pulses can phase shift circadian rhythms in humans and other animals. The problem is that it is not clear what type of phase dysfunction characterizes SAD patients. While one group of researchers claims that depression is associated with a phase advance of circadian rhythms,[13] another group claims that SAD is associated with a phase delay.[54] Of course, acute depression could involve a mechanism opposite to that of seasonal depression, but a third group of researchers found that SAD may not involve either phase advances or delays.[58] It would seem justified to conclude that the mechanism by which phototherapy relieves SAD symptoms is not yet known.

Whether or not the therapeutic action of bright light stimulation is known, the treatment itself may be effective. A major problem involving efficacy evaluation in the treatment of SAD is that well-controlled, double-blind studies (i.e., studies in which neither the patient nor the investigator can subjectively bias the results) are not feasible when light treatment is involved, as the patient can always tell whether light is administered or not.[19] The average treatment efficacy of phototherapy is reportedly about 50%, which is only slightly higher than the high end of the placebo effect for antidepressant drugs (20 to 40%).[19] A follow-up of SAD patients for almost nine years revealed a persistence of symptoms in 86% of the patients, which means that only 14% of the patients overcame the disease, even though 41% of them continued to use phototherapy (most of them in combination with antidepressant drugs).[59] Clearly, the outcome of treatment for SAD is not impressive. A recent placebo-controlled study revealed no difference between bright-light treatment and placebo in mean depression scores: patients receiving bright-light treatment improved just as much on average as patients receiving placebo.[60] An encouraging aspect of the results was that, after three to four weeks of treatment, more patients showed great improvement in the phototherapy group than in the placebo group. Much more research is still needed to settle the issue. For those patients who respond to the treatment, however, therapy already means the difference between endless miserable winters and an enjoyable productive life.

bibliographic notes

1. Spraycar, M. (1995). *PDR Medical Dictionary.* Montvale, NJ: Medical Economics.
2. Bunk, S. (1998). Ethical debate on placebo use may prompt new trial designs. *Scientist* 12(8): 1, 7, 14.
3. Bandura, A. (1969). *Principles of Behavior Modification.* New York: Holt, Rinehart and Winston.
4. Shapiro, S., Skinner, E. A., Kessler, L. G., von Korff, M., German, P. S., Tischler, G. L., Leaf, P. J., Benham, L., Cottler, L., and Regier, D. A. (1984). Utilization of health and mental health services. *Archives of General Psychiatry* 41: 971-978.
5. Eysenck, H. J. (1952). The effects of psychotherapy: An evaluation. *Journal of Consulting Psychology* 16: 319-324.
6. Levitt, E. E. (1957). The results of psychotherapy with children: An evaluation. *Journal of Consulting Psychology* 21: 189-196.
7. Astin, A. W. (1961). The functional autonomy of psychotherapy. *American Psychologist* 16: 75-78.
8. Smith, M. L., Glass, G. V., and Miller, T. I. (1980). *The Benefits of Psychotherapy.* Baltimore: Johns Hopkins University Press.
9. Shapiro, D. A. and Shapiro, D. (1982). Meta-analysis of comparative therapy outcome studies: A replication and refinement. *Psychological Bulletin* 92: 581-604.
10. Demeza, K. F. (1998). Getting personal: The business consulting approach to solving personal problems. *The Marketeer: Birmingham Business Today,* November 1998, p. 7.

11. Szaz, T.S. (1960). The myth of mental illness. *American Psychologist* 15: 113-118.
12. Foucault, M. (1972). *Histoire de la Folie à l'Âge Classique.* Paris: Gallimard.
13. Goodwin, F. K., Wirz-Justice, A., and Wehr, T. A. (1982). Evidence that the pathophysiology of depression and the mechanism of action of antidepressant drugs both involve alterations in circadian rhythms. In: Costa, E. and Racagni, G. (Eds.). *Typical and Atypical Antidepressants: Clinical Practice.* New York: Raven, pp. 1-11.
14. Kripke, D. F., Mullaney, D. J., Atkinson, M., and Wolf, S. (1978). Circadian rhythm disorders in manic-depressives. *Biological Psychiatry* 13: 335-351.
15. Szuba, M. P., Yager, A., Guze, B. H., Allen, E. M., and Baxter, L. R. (1992). Disruption of social circadian rhythms in major depression: A preliminary report. *Psychiatry Research* 42: 221-230.
16. Healy, D. and Waterhouse, J. M. (1995). The circadian system and the therapeutics of the affective disorders. *Pharmacology and Therapeutics* 65: 241-263.
17. Ehlers, C. L., Frank, E., and Kupfer, D. J. (1988). Social zeitgebers and biological timing. *Archives of General Psychiatry* 45: 948-952.
18. Lykken, D. and Tellegen, A. (1996). Happiness is a stochastic phenomenon. *Psychological Science* 7: 186-189.
19. Terman, M. (1988). On the question of mechanism in phototherapy for seasonal affective disorder: Considerations of clinical efficacy and epidemiology. *Journal of Biological Rhythms* 3: 155-172.
20. Nagayama, H., Hasama, N., Tsuchiyama, K., Yamada, K., Ihara, J., and Yanagisawa, T. (1992). Circadian temperature and sleep-wake rhythms in depression. *Japanese Journal of Psychiatry and Neurology* 46: 244-245.
21. Wehr, T. A. and Wirz-Justice, A. (1982). Circadian rhythm mechanisms in affective illness and in antidepressant drug action. *Pharmacopsychiatry* 15: 31-39.
22. Nikitopoulou, G. and Crammer, J. L. (1976). Change in diurnal temperature rhythm in manic-depressive illness. *British Medical Journal* 1976(1): 1311-1314.
23. Taillard, J., Sanchez, P., Lemoine, P., and Mouret, J. (1990). Heart rate circadian rhythm as a biological marker of desynchronization in major depression: A methodological and preliminary report. *Chronobiology International* 7: 305-316.
24. Stampfer, H. G. (1998). The relationship between psychiatric illness and the circadian pattern of heart rate. *Australian and New Zealand Journal of Psychiatry* 32: 187-198.
25. Tsujimoto, T., Yamada, N., Shimoda, K., Hanada, K, and Takahashi, S. (1990). Circadian rhythms in depression. II. Circadian rhythms in inpatients with various mental disorders. *Journal of Affective Disorders* 18: 199-210.
26. Willner, P. (9184). The validity of animal models of depression. *Psychopharmacology* 83: 1-16.
27. Baltzer, V. and Weiskrantz, L. (1975). Antidepressant agents and reversal of diurnal activity in the rat. *Biological Psychiatry* 10: 199-209.
28. Aschoff, J. (1989). Circadian activity rhythms in hamsters and rats treated with imipramine in the drinking water. *Chronobiologia* 16: 9-20.
29. Refinetti, R. and Menaker, M. (1993). Effects of imipramine on circadian rhythms in the golden hamster. *Pharmacology Biochemistry and Behavior* 45: 27-33.
30. Kripke, D. F. and Wyborney, V. G. (1980). Lithium slows rat circadian activity rhythms. *Life Sciences* 26: 1319-1321.

31. McEachron, D. L., Kripke, D. F., Sharp, F. R., Lewy, A. J., and McClellan, D. E. (1985). Lithium effects on selected circadian rhythms in rats. *Brain Research Bulletin* 15: 347-350.
32. Hafen, T. and Wollnik, F. (1994). Effect of lithium carbonate on activity level and circadian period in different strains of rats. *Pharmacology Biochemistry and Behavior* 49: 975-983.
33. Nagayama, H. (1996). Chronic administration of imipramine and lithium changes the phase-angle relationship between the activity and core body temperature circadian rhythms in rats. *Chronobiology International* 13: 251-259.
34. Possidente, B., Lumia, A. R., McEldowney, S., and Rapp, M. (1992). Fluoxetine shortens circadian period for wheel running activity in mice. *Brain Research Bulletin* 28: 629-631.
35. Duncan, W. C., Tamarkin, L., Sokolove, P. G., and Wehr, T. A. (1988). Chronic clorgyline treatment of Syrian hamsters: An analysis of effects on the circadian pacemaker. *Journal of Biological Rhythms* 3: 305-322.
36. Hallonquist, J., Lindegger, M., and Mrosovsky, N. (1994). Rubidium chloride fuses split circadian activity rhythms in hamsters housed in bright constant light. *Chronobiology International* 11: 65-71.
37. Marshall, E. (1994). Highs and lows on the research roller coaster. *Science* 264: 1693-1695.
38. Holden, C. (1994). A cautionary genetic tale: The sobering story of D2. *Science* 264: 1696-1697.
39. Andreasen, N. C., Arndt, S., Swayze, V., Cizadlo, T., Flaum, M., O'Leary, D., Ehrhardt, J. C., and Yuh, W. T. C. (1994). Thalamic abnormalities in schizophrenia visualized through magnetic resonance image averaging. *Science* 266: 294-298.
40. Benes, F. M. (1995). Is there a neuroanatomic basis for schizophrenia? An old question revisited. *Neuroscientist* 1: 104-115.
41. Schildkraut, J. J. (1965). The catecholamine hypothesis of affective disorders: A review of supporting evidence. *American Journal of Psychiatry* 122: 509-522.
42. Kramer, M. S., Cutler, N., Feighner, J., Shrivastava, R., Carman, J., Sramek, J. J., Reines, S. A., Liu, G., Snavely, D., Wyatt-Knowles, E., Hale, J. J., Mills, S. G., MacCoss, M., Swain, C. J., Harrison, T., Hill, R. G., Hefti, F., Scolnick, E. M., Cascieri, M. A., Chicchi, G. G., Sadowski, S., Williams, A. R., Hewson, L., Smith, D., Carlson, E. J., Hargreaves, R. J., and Rupniak, N. M. (1998). Distinct mechanism for antidepressant activity by blockade of central substance P receptors. *Science* 281: 1640-1645.
43. Abrams, R. (1989). Out of the blue: The rehabilitation of electroconvulsive therapy. *The Sciences* 29(6): 24-30.
44. Jacobsen, F. M., Wehr, T. A., Sack, D. A., James, S. P., and Rosenthal, N. E. (1987). Seasonal affective disorder: A review of the syndrome and its public health implications. *American Journal of Public Health* 77: 57-60.
45. Chung, Y. S. and Daghestani, A. N. (1989). Seasonal affective disorder: Shedding light on a dark subject. *Postgraduate Medicine* 86: 309-314.
46. Albert, P. S., Rosen, L. N., Alexander, J. R., and Rosenthal, N. E. (1991). Effect of daily variation in weather and sleep on seasonal affective disorder. *Psychiatry Research* 36: 51-63.

47. Kasper, S., Rogers, S. L. B., Yancey, A., Schulz, P. M., Skwerer, R. G., and Rosenthal, N. E. (1989). Phototherapy in individuals with and without subsyndromal seasonal affective disorder. *Archives of General Psychiatry* 46: 837-844.

48. Rosenthal, N. E., Levendosky, A. A., Skwerer, R. G., Joseph-Vanderpool, J. R., Kelly, K. A., Hardin, T., Kasper, S., DellaBella, P., and Wehr, T. A. (1990). Effects of light treatment on core body temperature in seasonal affective disorder. *Biological Psychiatry* 27: 39-50.

49. Magnusson, A. and Kristbjarnarson, H. (1991). Treatment of seasonal affective disorder with high-intensity light: A phototherapy study with an Icelandic group of patients. *Journal of Affective Disorders* 21: 141-147.

50. Eastman, C. I., Gallo, L. C., Lahmeyer, H. W., and Fogg, L. F. (1993). The circadian rhythm of temperature during light treatment for winter depression. *Biological Psychiatry* 34: 210-220.

51. Stewart, J. W., Quitkin, F. M., Terman, M., and Terman, J. S. (1990). Is seasonal affective disorder a variant of atypical depression? Differential response to light therapy. *Psychiatry Research* 33: 121-128.

52. Avery, D. H., Khan, A., Dager, S. R., Cox, G. B., and Dunner, D. L. (1990). Bright light treatment of winter depression: Morning versus evening light. *Acta Psychiatrica Scandinavica* 82: 335-338.

53. Wehr, T. A., Jacobsen, F. M., Sack, D. A., Arendt, J., Tamarkin, L., and Rosenthal, N. E. (1986). Phototherapy of seasonal affective disorder: Time of day and suppression of melatonin are not critical for antidepressant effects. *Archives of General Psychiatry* 43: 870-875.

54. Lewy, A. J., Sack, R. L., Miller, S., and Hoban, T. M. (1987). Antidepressant and circadian phase-shifting effects of light. *Science* 235: 352-354.

55. Jacobsen, F. M., Wehr, T. A., Skwerer, R. A., Sack, D. A., and Rosenthal, N. E. (1987). Morning versus midday phototherapy of seasonal affective disorder. *American Journal of Psychiatry* 144: 1301-1305.

56. Ruhrmann, S., Kasper, S., Hawellek, B., Martinez, B., Höflich, G., Nickelsen, T., and Möller, H. J. (1998). Effects of fluoxetine versus bright light in the treatment of seasonal affective disorder. *Psychological Medicine* 28: 923-933.

57. Terman, M., Terman, J. S., and Ross, D. C. (1998). A controlled trial of timed bright light and negative air ionization for treatment of winter depression. *Archives of General Psychiatry* 55: 875-882.

58. Levendosky, A.A., Joseph-Vanderpool, J.R., Hardin, T., Sorek, E., and Rosenthal, N.E. (1991). Core body temperature in patients with seasonal affective disorder and normal controls in summer and winter. *Biological Psychiatry* 29: 524-534.

59. Schwartz, P. J., Brown, C., Wehr, T. A., and Rosenthal, N.E. (1996). Winter seasonal affective disorder: A follow-up study of the first 59 patients of the National Institute of Mental Health seasonal studies program. *American Journal of Psychiatry* 153: 1028-1036.

60. Eastman, C.I., Young, M.A., Fogg, L.F., Liu, L., and Meaden, P.M. (1998). Bright light treatment of winter depression: A placebo-controlled trial. *Archives of General Psychiatry* 55: 883-889.

chapter fourteen

other applications

> Basic knowledge about circadian rhythms can be used
> to enhance the therapeutic action of medicines and to
> optimize the planning of astronaut schedules for space
> exploration.

In the preceding three chapters, we saw how knowledge about basic mechanisms of circadian rhythmicity can be applied to real-life situations such as jet lag, shift work, sleep disorders, depression, and seasonal affective disorder. In this final chapter, we will examine additional applications in chronopharmacology and space exploration.

chronopharmacology

We have seen that circadian rhythmicity affects numerous physiological and behavioral processes. Examples in addition to those previously mentioned include daily oscillations in visual sensitivity[1] and pain tolerance.[2] Therefore, it is not surprising to find out that the symptoms of chronic diseases — such as allergic rhinitis, angina, arthritis, asthma, epilepsy, hypertension, and ulcer disease — exhibit circadian rhythmicity.[3] Awareness of these rhythmic patterns is essential for proper diagnosis of the disease.

Recognition of the existence of circadian rhythms is important not only for diagnostic purposes but also for therapeutic purposes. Indeed, if physiological functions oscillate daily, it is to be expected that the sensitivity of the patient to therapeutic agents will also oscillate daily. Medical research has clearly demonstrated that the pharmacodynamics and pharmacokinetics of H2-blockers, antiasthmatics (theophylline, beta-agonists, anticholinergics, glucorticoids), and cardiovascular active drugs (beta blockers, organic nitrates, calcium channel blockers), to name a few, display daily rhythmicity.[4] The treatment of hypertension is an excellent example of the importance of chronopharmacology (i.e., the administration of drugs at specific times of the day when the drugs are most effective and have least side effects). Blood pressure has a definite and reproducible circadian pattern: it is highest during the day, lowest during sleep, and increases rapidly from 4 am to noon. Consequently, antihypertensive control should be concentrated in the early morning. In order to attain peak plasma level of the medication during the early morning period, controlled-onset, extended-release delivery systems have been developed and implemented. Such systems provide effective control of hypertension by supplying the antihypertensive drug at the time when blood pressure is highest.[5]

The magnitude of allergic responses, like that of all other physiological processes, is expected to oscillate daily. While several clinical studies have failed to demonstrate daily rhythmicity in skin reactivity to allergens,[6–8] much evidence has been accumulated in support of the rhythmic nature of asthmatic symptoms.[9,10] The first large-scale investigation of daily rhythmicity in asthma was that of Turner-Warwick involving 7,729 asthmatic outpatients.[11] Turner-Warwick confirmed previous anecdotal evidence of a worsening of symptoms during the late night and early morning. The "morning dip" in expiratory flow is truly an organic process and not a mere consequence of the lack of effective medication during the night.[12] However, controlled laboratory studies have not been conducted to exclude the role of environmental factors (such as changes in quantity and quality of allergens) in the determination of the daily oscillation. Thus, in strict terms, one can say that asthmatic symptoms exhibit daily rhythmicity but not necessarily *circadian* rhythmicity.

Two indices of expiratory flow have been used as objective measures to facilitate the management of asthma: peak expiratory flow rate (PEF)[12–15] and forced expiratory volume in 1 second (FEV1).[16–18] The two measures correlate well with each other and with the caliber of the airway.[19] Measurements conducted at several time points during a day revealed clear rhythmicity in asthmatic patients as well as in normal subjects.[9,10] For both groups, peak lung function occurs at approximately 4 pm and lowest lung function at 4 am. The two groups differ in the mean level and amplitude of the oscillation: asthmatic patients display lower lung function than normal subjects overall, and their "morning dip" is more accentuated than that of normal subjects.

In a study by Karras and colleagues,[20] asthmatic patients who presented nocturnally to the emergency department did not differ from other asthmatic patients in disease severity. The authors interpreted this finding as an indication that daily rhythmicity in asthma severity is not clinically relevant for purposes of emergency management. This interpretation is correct in strict sense, but one should not infer from this that daily rhythmicity in asthma severity is not clinically relevant in general. Since a large number of uncontrolled variables determines a patient's decision to leave home in the middle of the night and go to a hospital emergency department, the study has a natural sampling bias that prevents any generalization to the whole population of asthmatic patients.

Chronobiological treatment of asthma involves the same drugs used in traditional approaches (corticosteroids, theophylline, anticholinergics, or beta-adrenergic agonists) but makes use of chronobiological information to maximize pharmacological effects and minimize side effects.[10,21] Since the greatest extent of bronchial inflammation is observed in the early morning, one would expect the best timing for drug administration to be the early evening (to allow enough time for drug action). Indeed, drug administration in the early evening has been found to be optimal for corticosteroids,[22] theophylline,[23] anticholinergics,[24] and beta-adrenergic agonists.[25] While this approach does not provide new clues regarding an eventual cure for asthma, it does improve the patient's condition and offers a more effective way to prevent asthmatic episodes.

space exploration

Cosmonaut Yuri Gagarin's orbital flight in 1961 and astronaut Neil Armstrong's landing on the moon in 1969 were rightfully heralded as living proof of the possibility of space exploration. The Russian space station and the American space shuttle programs added support to the idea. The worldwide-web site of the National Aeronautics and Space Administration (www.nasa.gov) proudly states that the goals of its Human Exploration and Development of Space enterprise is to "increase human knowledge of nature's processes using the space environment, explore the solar system, achieve routine space travel, and enrich life on earth through people living and working in space." Colonization of other planets is a more remote but not unlikely possibility.

Aside from the complexity of new developments in spacecraft design and operation, such as onboard computers capable of assuming complete control of the spacecraft,[26] what type of challenges can we expect? How will humans adapt to life away from Earth? Of course, this is a very complex question, but we can limit it here to aspects dealing with circadian rhythmicity. Hard data about human physiology in space are not plentiful, mainly because participation in basic research deprives astronauts and cosmonauts of valuable time needed to accomplish their multiple scheduled tasks. Nevertheless, some information on biological rhythms has been collected since the 1960s.[27-30] Although reductions in sleep time and changes in sleep pattern (as compared to pre-flight baseline measurements) have been observed, alterations in circadian rhythms have been found to be minimal. In one study aboard the Russian station MIR, a 2-hour delay was noticed in the body temperature rhythm,[29] although the shift could have been due to a shift in the work schedule. In a study aboard the American space shuttle Columbia, where disruptions in the work schedule were minimized, no significant changes in the amplitude and phase of the rhythms of body temperature, urinary melatonin sulfate, and free cortisol were observed.[30] In neither of the studies (which lasted 30 and 17 days, respectively) were the rhythms found to free-run. In the Columbia study, the astronauts slept in light- and sound-shielded bunks with no light during the sleep period and spent the day performing their duties under artificial illumination, which was less bright than that experienced during the baseline period on earth but, it would seem, was sufficient to maintain entrainment.

If it is true that the very reduced gravity and the lack of other geophysical influences do not have a great impact on circadian rhythms, then the major concern about space exploration will be about rhythm entrainment. Apparently, the artificial living conditions maintained in spacecrafts are sufficient to preserve entrainment. However, artificial conditions are not viable in large scale enterprises. Every year, we advance our clocks one hour in the spring and turn them back in the fall to save millions of dollars in electricity. If we had to maintain a full light–dark cycle artificially, the costs would be exhorbitant. So, if we are ever to colonize a planet whose day is longer or shorter than 24 hours, we had better learn to adapt our circadian systems to non-24-hour days. Our earthly biological clock can be entrained to light–dark cycles that do not differ much from 24 hours. But what if we want to colonize a planet that has a 35-hour day? In some cases, frequency demultiplication (discussed in Chapter 5) could help us deal with unusual daylengths (say, 48-hour days), but this will not be the case for planets with 15-hour days or 35-hour days. A better understanding of entrainment mechanisms might lead to the development of potent and inexpensive zeitgebers capable of entraining us to the unusual daylength, although we would probably still face the problem of wanting to be awake for much of the long alien night. As we learn more about the mechanisms of action of circadian genes, biological engineering of a circadian mutant gene for humans (that could set our circadian period to that of our destination planet) may provide the optimal solution.

Before we colonize an alien planet, we must get there. As much as we may increase spacecraft speed, we will be limited to the speed of light, which means that voyages may take not weeks but generations. Assuming that cryopreservation is not an option, travelers will likely spend their whole lives in space. The costs of maintaining an appropriate light–dark cycle will probably be negligible in comparison with the costs of generating breathable air and providing food, neutral ambient temperature, and health care for all the travelers. Nevertheless, accurate knowledge of human circadian rhythms and the zeitgebers that affect them will be essential to provide the sojourners with an environment that relieves rather than accentuates the stress of the journey.

While basic research on circadian rhythms proceeds on earth, occasional studies will be conducted aboard the International Space Station, whose on-orbit assembly started in 1998 and is scheduled for completion in 2004. The largest scientific and technological endeavor ever undertaken by humankind, the station is being built by the United States in collaboration with Canada, Japan, Russia, Brazil, Belgium, Denmark, France, Germany, Italy, Holland, Norway, Spain, Sweden, Switzerland, and the United Kingdom. The program currently involves more than 100,000 people at space agencies and contractor companies around the world. The station will provide a permanent laboratory for research and development in multiple fields, from materials science to human physiology.

bibliographic notes

1. Bassi, C. J. and Powers, M. K. (1986). Daily fluctuations in the detectability of dim lights by humans. *Physiology and Behavior* 38: 871-877.
2. Hildebrandt, G., Strempel, H., and Will, H. D. (1992).Circadian variations of the subjective habituation to cold pain. *Journal of Interdisciplinary Cycle Research* 23: 222-224.
3. Smolensky, M. H. and D'Alonzo, G. E. (1993). Medical chronobiology: Concepts and applications. *American Review of Respiratory Diseases* 147(6): S2-S19.
4. Lemmer, B. (1996). The clinical relevance of chronopharmacology in therapeutics. *Pharmacological Research* 33: 107-115.
5. Neutel, J. M. and Smith, D. H. (1997). The circadian pattern of blood pressure: Cardiovascular risk and therapeutic opportunities. *Current Opinions in Nephrology and Hypertension* 6: 250-256.
6. Rehn, D., Geissler, H., Schonbrunn, U., Lukas, H., Rahlfs, V. W., and Hennings, G. (1990). Variations of skin sensitivity to intracutaneous histamine provocations with regard to provocation time. *Arzneimittelforschung* 40: 777-781.
7. Paquet, F., Boulet, L. P., Bedard, G., Tremblay, G., and Cormier, Y. (1991). Influence of time of administration on allergic skin prick tests response. *Annals of Allergy* 67(2): 163-166.
8. Vichyanond, P. and Nelson, H.S. (1989). Circadian variation of skin reactivity and allergy skin tests. *Journal of Allergy and Clinical Immunology* 83: 1101-1106.
9. Pincus, D. J., Beam, W. R., and Martin, R. J. (1995). Chronobiology and chronotherapy of asthma. *Clinics in Chest Medicine* 16: 699-713.
10. Di Stefano, A., Lusuardi, M., Braghiroli, A., and Donner, C. F. (1997). Nocturnal asthma: Mechanisms and therapy. *Lung* 175: 53-61.
11. Turner-Warwick, M. (1988). Epidemiology of nocturnal asthma. *American Journal of Medicine* 85(1B): 6-8.
12. Dethlefsen, U. and Repgas, R. (1985). Ein neues Therapieprinzip bei nachtlichen Asthma. *Klinische Medizin* 80: 44-47.
13. Lebowitz, M. D., Krzyzanowski, M., Quackenboss, J. J., and O'Rourke, M. K.(1997). Diurnal variation of PEF and its use in epidemiological studies. *European Respiratory Journal* 24: 49S-56S.
14. Iwasaki, Y., Ueda, M., Hashimoto, S., Hashikura, H., Mizobuchi, K., Arimoto, T., Itoh, H. Hara, H., and Nakagawa, M. (1996). Optimal time of the day for measuring peak expiratory flow rates in patients with asthma. *Nippon Kyobu Shikkan Gakkai Zasshi* 34: 885-889.
15. Meijer, G. G., Postma, D. S., van der Heide, S., de Reus, D. M., Koeter, G. H., Roorda, R. J., and van Aalderen, W. M. (1996). Seasonal variations in house dust mite influence the circadian peak expiratory flow amplitude. *American Journal of Respiratory Critical Care Medicine* 154: 881-884.
16. Kondo, S. and Ito, M. (1997). Reevaluation of the relationship between within-day bronchial variability and bronchial responsiveness in asthmatic children. *Chronobiology International* 14: 61-69.

17. Dottorini, M. L., Tantucci, C., Peccini, F., Grassi, V., and Sorbini, C. A. (1996). Diurnal change of bronchial caliber and airway responsiveness in asthmatics during long-term treatment with long-acting beta 2 agonist salmeterol. *International Journal of Clinical Pharmacology and Therapeutics* 34: 438-443.

18. Kraft, M., Wenzel, S. E., Bettinger, C. M., and Martin, R. J. (1997). The effect of salmeterol on nocturnal symptoms, airway function, and inflammation in asthma. *Chest* 111: 1249-1254.

19. Troyanov, S., Ghezzo, H., Cartier, A., and Malo, J. L. (1994). Comparison of circadian variations using FEV1 and peak expiratory flow rates among normal and asthmatic subjects. *Thorax* 49: 775-780.

20. Karras, D. J., D'Alonzo, G. E., and Heilpern, K. L. (1995). Is circadian variation in asthma severity relevant in the emergency department? *Annals of Emergency Medicine* 26: 558-562.

21. Burioka, N. and Sasaki, T. (1996). Chronopharmacology and chronotherapy for asthma by using PEF. *Nippon Rinsho* 54: 2956-2961.

22. Reinberg, A., Gervais, P., Chaussade, M., Fraboulet, G., and Duburque, B. (1983). Circadian changes in effectiveness of corticosteroids in eight patients with allergic asthma. *Journal of Allergy and Clinical Immunology* 71: 425-433.

23. Reinberg, A., Pauchet, F., Ruff, F., Gervais, A., Smolensky, M. H., Levi, F., Gervais, P., Chaouat, D., Abella, M. L., and Zidani, R. (1987). Comparison of once-daily evening versus morning sustained release theophylline dosing for nocturnal asthma. *Chronobiology International* 4: 409-419.

24. Morrison, J., Pearson, S., and Dean, H. (1988). Parasympathetic nervous system in nocturnal asthma. *British Medical Journal* 296: 1427-1429.

25. Gaultier, L., Reinberg, A., and Motohashi, Y. (1988). Circadian rhythm in total pulmonary resistance of asthmatic children: Effects of a beta-agonist agent. *Chronobiology International* 5: 285-290.

26. Normile, D. (1998). NASA craft to take the controls in flight. *Science* 282: 604-605.

27. Adey, W. A., Kado, R. T., and Walter, D. O. (1967). Computer analysis of EEG data from Gemini flight GT-7. *Aerospace Medicine* 38: 345-359.

28. Frost, J. D., Shumate, W. H., Booher, C. R., and DeLucchi, M. R. (1975). The Skylab sleep monitoring experiment: Methodology and initial results. *Acta Astronautica* 2: 319-336.

29. Gundel, A., Polyakov, V. V., and Zulley, J. (1997). The alteration of human sleep and circadian rhythms during spaceflight. *Journal of Sleep Research* 6: 1-8.

30. Monk, T. H., Buysse, D. J., Billy, B. D., Kennedy, K. S., and Willrich, L. M. (1998). Sleep and circadian rhythms in four orbiting astronauts. *Journal of Biological Rhythms* 13: 188-201.

part V

---

appendix

appendix

doing it yourself: software for data analysis and simulation

Several computer programs especially designed for circadian physiology can be found in the disk/CD that accompanies this book. The programs allow the reader to acquire practical experience in analyzing data and simulating circadian processes and, therefore, to enhance the understanding of the material presented in the preceding chapters. Although very simple to use, the programs are powerful enough to be used also by researchers interested in the analysis of their own experimental data.

While readers are welcome to use the programs as they see fit, it is recommended that the procedures described in this section be followed first. Reading at least Chapters 4 through 7 before using the software is strongly advised.

installing the software

Minimal hardware is required to install and use the programs. Any version of Microsoft Windows from Windows 3.0 on (including Windows 95, Windows 98, Windows NT, and Windows 2000) is appropriate. A little over one megabyte of free hard disk space and no more than two megabytes of RAM are required. For technical support, you may send e-mail questions to the author at help@circadian.org.

To install the software, insert the disk/CD into the appropriate drive and run the setup program (SETUP.EXE) by clicking on its icon in File Manager, Windows Explorer, or My Computer (or using the Run option in the Taskbar). You may be asked to confirm the name of the folder in which the programs will be installed (the default is Bioclock). After the programs are installed, you can access them through File Manager, Windows Explorer, or My Computer. If you have Windows 95 or above, a shortcut to the Bioclock folder will be automatically placed in the Taskbar (Start Menu).

inspecting and analyzing data

exercise 1: cartesian plots

The first step in any data analysis procedure is to casually inspect the data. The simplest way to inspect data on circadian rhythms is to plot them in standard Cartesian format (time on the abscissa and experimental variable on the ordinate). You can use your own data in this exercise, as long as they are arranged as a single stream of data points in an ASCII (text) file. If you have collected data manually, you can prepare a data file with your word processor (and save it in text format). First, however, you should practice with the data files provided with this software.

1. Start the program **Plot.exe**.
2. In the *Source* panel, select the sub-folder *Data* by double-clicking on it.
3. Select the first data file, **A01.dat**, by clicking on it. This file contains the records of body temperature of a Richardson's ground squirrel measured by telemetry every six minutes for a week. Since the data points are equally spaced in time, and we are not concerned about the absolute time of the day at this point, the file contains only values for the ordinate.
4. The *Data* panel allows you to indicate the appropriate bin size, number of bins to be skipped (if any), the number of bins to be used (files shorter than the specified length are loaded in their entirety), and optional filters. For this exercise, none of the default values need to be changed. Of course, the choice of bin size must match the bin size used at data collection. In my laboratory, I normally use 6-min bins because they correspond to an exact decimal fraction of an hour (i.e., 0.1 h).
5. Click on *Cartesian* to plot the data. The first day of the recordings will be shown, and you can use the scroll bar to scroll through the whole data set. Notice that, on each day, body temperature rises after approximately 12 o'clock.
6. If you wish to have a printout of any single day, click on *Print*.
7. When done inspecting this data set, select the data file **A02.dat** by clicking on it. This file contains the records of body temperature of a degu (a South American rodent) measured by telemetry every six minutes for eight consecutive days.
8. As was the case for the previous data set, the default values in the *Data* panel need not be changed. Click on *Cartesian*.
9. You will notice that the data set contains many spurious readings that disturb the inspection of the measurements. This is a consequence of poor data collection.
10. There are two ways by which we can improve the visual display. One of them is to simply click on the *Dots* button (right under the scroll

bar). Browse through the eight days and notice that the spurious data points are not as disruptive this way.

11. A second way to improve the visual display is to filter the data. Click on the *Lines* button (to restore the line plot) and then on the *Low* filter box. Type 36 and then click on the *High* filter box. Type 39 and then click on *Cartesian*. Browse again. The data still do not look very good, but they look better than without the filters.

exercise 2: actograms

The actogram is a classic graphic in circadian physiology. Originally used only for records of running-wheel activity, it is now used for practically any type of variable that is recorded over an extended period of time. You can use your own data in this exercise, as long as they are arranged as a single stream of data points in an ASCII (text) file. First, however, you should practice with the data files provided with this software.

1. Start the program **Plot.exe.**
2. In the *Source* panel, select the sub-folder *Data* by double-clicking on it.
3. Select the data file **A03.dat** by clicking on it. This file contains the records of running-wheel activity of a golden hamster maintained in constant darkness for 36 consecutive days. The number of wheel revolutions is accumulated into 6-min bins (for a total of 8,640 data points in the file). As in Exercise 1, the file contains only values for the ordinate.
4. For now, leave the default values in the *Data* panel. Click on *Actogram* to display the data. You can see why golden hamsters are the preferred rodent for the study of circadian rhythms.The pattern of activity is very "clean," with wheel-running neatly restricted to a limited portion of each day. You can also notice that the onsets of activity are neatly arranged one under the other in almost a vertical line, which indicates a free-running period very close to 24.0 h.
5. Select the data file **A04.dat** by clicking on it. This file contains the records of running-wheel activity of another hamster maintained in constant darkness for 29 days.
6. Click on *Actogram* to display the data. The onsets for this animal clearly deviate from a vertical line, indicating a free-running period slightly longer than 24.0 h. We will see how to determine the exact period in Exercises 4 and 5.
7. Select the data set **A05.dat**. This file contains the records of body temperature of a Long-Evans rat measured by telemetry every 6 min for 6 weeks. A light-dark cycle was present for the first 4 weeks.
8. Click on *Actogram*. What do you see? Probably 42 horizontal straight lines. Why? Because body temperature, unlike locomotor activity, does not go down to zero during the inactive phase of the circadian cycle. By plotting every value above zero, we end up plotting every

single data point. Thus, in order to have a useful actogram, we need to clip off the lower values. An arbitrary but convenient clipping level is the mesor of the rhythm (the mean body temperature for the whole data set). We will see later how to calculate the mesor. For now, click on the *Clip* box and type 36.2.

9. Click on *Actogram*. What a difference! Now you have a very legible actogram of the rat's body temperature rhythm. You can clearly see that the animal exhibited a period of 24.0 h during the two weeks under a light-dark cycle and that it free-ran with a period longer than 24.0 h when released into constant darkness.

10. Select the data set **A06.dat**. This file contains the records of locomotor activity of a pill bug (a small terrestrial crustacean) measured with an infrared photocell for 19 days in constant darkness. As usual, the data resolution is 6 min, and the file contains only the ordinate values.

11. Click on the *Clip* box, delete 36.2, and type 0.

12. Click on *Actogram*. You can see why pill bugs are *not* the preferred species in circadian physiology. The records are much "noisier" than those of the hamsters. Nonetheless, you will notice that the animal free-runs with a period much shorter than 24.0 h.

exercise 3: calculating the mesor and amplitude of a rhythm

Circadian rhythms are defined by their mean level (mesor), amplitude, period, phase, and waveform. In this exercise, we will deal with the analysis of the first two parameters.

1. Start the program **Plot.exe**.

2. In the *Source* panel, select the sub-folder *Data* by double-clicking on it.

3. Select the data file **A05.dat** by clicking on it. As mentioned in the previous exercise, this file contains the records of body temperature of a Long-Evans rat measured by telemetry every 6 min for six weeks, a light–dark cycle was present for the first four weeks, and only the ordinate values are included.

4. Since different environmental conditions prevailed during the last two weeks, let us restrict the analysis to the first four weeks. To do so, click on the scroll bar for *Bins to use* to bring the value down from 14,400 to 6,720 (once you get to 6,700, use the right scroll button to add the final 20 bins).

5. Click on *Cartesian*. While loading the data, the program calculates the arithmetic mean of the 6,720 values and determines the lowest and highest values in the data set. They are shown above the graph. In this case, the mean (mesor) is 36.142°C, as shown also by the horizontal blue line. The lowest value is 34.92°C and the highest value is 37.53°C. As discussed in Chapter 7, many (but not all) authors refer to the amplitude of a circadian rhythm as the difference between the highest and the lowest values. In this case, the amplitude is 2.61°C.

6. It should be mentioned that the highest and lowest values will be reduced if narrower filters are used. For example, if you set the *Low* filter to 35.2 and the *High* filter to 37 and then click on *Cartesian* again, the amplitude of the rhythm will be reduced to 1.8°C. To prevent filter manipulations from affecting the calculation of circadian parameters, several procedures have been developed to avoid the use of filters, such as choosing the minimal value as the lowest value recorded at least 10 times during 10 days,[1] making use of frequency histograms,[2] or using the range of values of a smooth curve interpolated to the data rather than the range of values of the actual data.[3]

7. Select the data file **A07.dat** by clicking on it. This file contains the records of metabolic heat production (in watts) of a fat-tailed gerbil measured by indirect calorimetry every 6 min for six consecutive days.

8. If you changed the default filter values, set *Low* back to 0 and *High* back to 1000.

9. Click on *Cartesian*. Here we have a mean of 1.071 W and an amplitude of 2.08 W. As you browse through the six days, you will notice that the heat production values oscillate around the mean (blue line). For this nocturnal animal, higher values of heat production are observed during the dark phase of the light–dark cycle (the first 12 h of each day).

exercise 4:   calculating circadian period

Having dealt with the mesor and amplitude of circadian rhythms in the preceding exercise, we will now deal with the issue of calculating the circadian period. Before we proceed to traditional methods of calculating circadian period (using **Wintau.exe**), we will employ a very simple, intuitive procedure (using the now familiar **Plot.exe**).

1. Start the program **Plot.exe**.

2. In the *Source* panel, select the sub-folder *Data* by double-clicking on it.

3. Select the data file **A06.dat** by clicking on it. As mentioned in Exercise 2, this file contains the records of locomotor activity of a pill bug measured with an infrared photocell for 19 days in constant darkness. As usual, the data resolution is 6 min, and the file contains only the ordinate values.

4. Click on *Actogram*. Although the records are noisy, it is evident that the circadian period is shorter than 24.0 h. We know that it is shorter than 24.0 h because the daily onsets do not align along a vertical line. But what if a day were shorter than 24.0 h? If the day were as short as the circadian period, the onsets would align along a vertical line. Thus, if we can find out the daylength that causes the onsets to align along a vertical line, we will know what the circadian period is.

5. Click on the down arrow by the text "Modulo 24 h" several times. As the plot modulo (daylength) shortens, the onsets move toward a

vertical line. When you reach 23.3 h, they are almost perfectly aligned. Thus, the free-running period of this animal is 23.3 h.

6. Select the data file **A08.dat**. This file contains artificial data constructed as a series of cosine waves with a period of 23.5 h. Click on the *Clip* box and type 5 (which is the mesor of the rhythm). Then click on *Actogram* and observe the smooth actogram.

7. Select the data file **A09.dat**. This is also an artificial file, but 60% of the data points in the preceding file were replaced with random noise in the range of oscillation. Click on *Actogram* and observe the noisier but still clear rhythm.

8. Select the data file **A10.dat**. This file contains 85% noise. Click on *Actogram* and observe the unintelligible actogram. The period is still 23.5 h, but there is so much noise that you cannot see it. You need a better method to calculate circadian period when the data are noisy (which is very common in mammals with partial suprachiasmatic lesions, for instance). Also, of course, it would be helpful to have a method that calculates circadian period without requiring a human observer — not only to make the experimenter's life easier but also to avoid observer's bias.

9. Close **Plot.exe** and open **Wintau.exe**. Wintau calculates circadian period using the periodogram procedure developed by Enright [4] and refined by Dörrscheidt and Beck,[5] and Sokolove and Bushnell.[6] The approach taken by Sokolove and Bushnell utilizes the chi-square (χ^2) distribution for the determination of statistical significance and, accordingly, is referred to as the chi-square periodogram. Many other procedures have been developed or adapted to calculate circadian period, including Fourier analysis,[7] autocorrelation,[8] linear regression of onsets,[9] inter-onset averaging,[10] and acrophase counting.[11] Of these, the chi-square periodogram presents the best combination of accuracy, tolerance to waveform irregularity, and tolerance to noise.[12] With minor adjustments, the chi-square periodogram can also be used to calculate the period of other biological rhythms, such as the estrous cycle.[13]

10. In the *Source* panel, select the sub-folder *Data* by double-clicking on it.

11. Select the data file **A08.dat** (the one with 20 days of smooth cosine waves). Do not change the defaults in the *Data* and *Period* panels. Click on *OK*.

12. For each potential period between 20.0 and 26.0 h, a Q_p value is given. The higher the Q_p value, the closer to real is the period associated with it. In this case, the highest Q_p is associated with a period of 23.5 h (as it should be, since the data set was artificially constructed to have a period of 23.5 h). Of course, the highest Q_p in a periodogram may not be absolutely (statistically) high. This is where the chi-square test comes in. As shown at the top of the table, a Q_p of 2350 is significantly different from noise at a significance level better than 0.01. As a matter of fact, this statistical significance is rather conser-

vative. Since 61 periods are being tested simultaneously (i.e., 20.0 to 26.0 in steps of 0.1), the level of significance is adjusted upward. In the *Period* panel, click on the scroll bars for *Start at* and *End at* so as to set the range at 23.1 to 23.9 h. Then click *OK* again. The Q_P remains 2350, but the significance level is now better than 0.001. Thus, the shorter the range, the more sensitive is the test. Of course, you should always set the range of the periodogram based on an honest expectation. To maliciously reduce the range based on previous test is to defeat the purpose of statistical testing.

13. Set *Start at* back to 20 and *End at* back to 26. Select data file **A09.dat** (the one with 60% noise). Click on *OK*. The highest Q_P is now only 750 (down from 2350), and it is off by a decimal unit (23.6 h rather than 23.5 h). The level of significance is still better than 0.01, however.

14. Select data file **A10.dat** (the one with 85% noise). Click on *OK*. Do not be surprised that the best Q_P is associated with a period of 25.5 h. This happened by chance, as indicated by the non-significance indicated at the top of the table ($p > 0.05$). Is 85% noise too much noise? Not necessarily. Click on the *Bins to use* scroll bar in the *Data* panel. A window will open telling you that, as in any statistical test, you can improve the power of the chi-square periodogram by increasing the sample size. Keep clicking on the scroll bar until you reach 4800 (which is the maximal number of bins available in the file). This corresponds to 20 days rather than 10. Click on *OK*. Darn! The period estimate is now very close (23.4 h rather than 23.5 h), but it is still non-significant. But then, why are we using a range from 20 to 26 h? Is it not reasonable to use a range from 21 to 25 instead? Try it. You will be pleased.

exercise 5: calculating circadian period (continued)

As mentioned above, the chi-square periodogram procedure (used in **Wintau.exe**) presents the best combination of accuracy, tolerance to waveform irregularity, and tolerance to noise in the determination of circadian period. However, it assumes that you have a flawless data collection procedure. That is, your data files must contain equally spaced data points, with no missing points. If your data set has a few missing points, you may replace them with the value of the mesor; however, if many points are missing, or if the data were collected at irregular intervals, you cannot use **Wintau.exe**. In this case, you may use **Sloppy.exe**.

Quite recently, Dr. Thomas Ruf, from Vienna Veterinary University, demonstrated that a procedure used in astrophysical sciences — the Lomb–Scargle periodogram — can be used for the analysis of biological rhythms.[14] Although the procedure has not been thoroughly tested in biological research, it seems to surpass the chi-square periodogram in accuracy, tolerance to waveform irregularity, and tolerance to noise *in addition to* allowing the analysis of data sets with missing points or irregular sampling intervals.

On top of all this, the Lomb–Scargle periodogram does not falsely identify harmonics of the main oscillation as the chi-square periodogram does.[12,14]

1. Start the program **Sloppy.exe**. This program uses a graphic procedure to determine circadian period. It expects data files with two columns, the first one containing time stamps (hours in decimal format, from 0.0 at midnight to 23.9 at 11:54 pm) and the second one containing the values of the physiological variable. Circadian period is calculated by a non-parametric procedure. An index of rhythmicity (R) is calculated separately and tested for statistical significance.
2. Select the sub-folder *Data* by double-clicking on it.
3. Select the data file **A11.dat** by clicking on it. This file contains artificial data constructed as a series of cosine waves with a period of 23.5 h (unlike file **A08.dat**, this file contains time stamps).
4. Click on *OK*. You will see a plot of the raw data on the lower panel and a 36-h educed wave on the upper right panel.
5. After a few seconds (or minutes, if your computer is very slow), the upper right panel will show a 24-h educed wave and the results of the analysis. The calculated period is 23.4 h (only 0.1 h off the correct value) and R is 0.1 (which is associated with a probability of less than 0.01 by chance, which means that there is significant circadian rhythmicity).
6. Select the data file **A12.dat**. This is also an artificial file, but 60% of the data points in the preceding file were replaced with random noise in the range of oscillation. Click on *OK*. Observe the noisier but still clear rhythm on the lower panel and the not-so-clear 36-h educed wave. The results of the analysis indicate a period of 24.0 h (which is half-an-hour off the correct value) and an R value of 0.4 (which is still significant at the 0.01 level). The increased noise reduced, but did not eliminate, the ability of the program to accurately calculate the circadian period.
7. Select the data file **A13.dat** (which contains a cosine wave with 85% noise) and click on *OK*. The analysis produces a very unreliable estimate of period (29.4 h) but indicates that there is no significant rhythmicity, which means that the period estimate should be ignored. Thus, the increased noise eliminated the ability of the program to accurately calculate circadian period.
8. Select the data file **A14.dat**. This file contains records of body temperature of a laboratory rat with a few missing points (more exactly, 10% of the 6-min measurements over 7 days are missing). Click on *OK*. The analysis indicates a circadian period of 23.8 h (which is consistent with the apparent period inferred from the plot of the raw data) and an R value of 0.3 (which is significant at the 0.01 level).
9. Select the data file **A15.dat**. This file is similar to **A14.dat** but now 80% of the data points are missing. Click on *OK*. Not surprisingly in view of the large proportion of missing points, the analysis indicates

no significant rhythmicity (p > 0.05) and an unrealiable estimate of period (26.0 h).

10. To make sure that you understand the file structure expected by the program, close **Sloppy.exe** and open your word processor. Load the file **A15.dat** (which should be in the C:\Bioclock\Data folder or in the alternate folder that you chose when the circadian software was installed). As you can see, the structure of the file is rather simple. Each time point (in a 24-h clock) is paired with a measurement of body temperature.

exercise 6: calculating circadian acrophase

The previous exercises dealt with three of the five parameters of circadian rhythms (mesor, amplitude, and period). This exercise deals with the phase parameter. We will deal only with data files containing the ordinate values in this exercise. Manually entered data, as well as files containing time stamps, will be dealt with in Exercise 7. The fifth parameter (waveform) will not be addressed at all, for the simple reason that there is no standard procedure for analyzing waveform. Technically, any wave can be described by Fourier analysis as the sum of multiple sine and cosine waves;[7] however, a comparison of the waveform of different circadian rhythms by Fourier analysis has never been conducted and, consequently, it is not known whether it would be of actual physiological relevance.

1. Start the program **Acro.exe**. This program calculates the acrophase of a rhythm (the time at which the rhythm peaks each day). Of course, real rhythms do not have perfect waveforms and, consequently, do not peak at exact intervals. Still, if the period is constant and the waveform is not totally random, we should be able to identify an average acrophase. We could do it manually by smoothing the rhythm and visually determining the acrophases.[15] But we can also determine the acrophase by fitting a mathematically defined curve to the data and determining the acrophase of the fitted curve,[3,16,17] thus avoiding observer's bias. This program fits a cosine wave to the data to determine the acrophase.

2. Choose *Load data from disk*.

3. Select the sub-folder *Data* by double-clicking on it.

4. Select the data file **A16.dat** by clicking on it. This file contains the records of body temperature of a fat-tailed gerbil measured by telemetry every six minutes for a week under a 24-h light–dark cycle. As usual, the file contains only values for the ordinate.

5. Click on *OK* (or double-click on the file name).

6. Click on *File contains data only*.

7. Since the data were collected every 6 min, the resolution of the data is 0.1 h, which is the default value. Click on *OK*.

8. Since the animal was under a 24-h light-dark cycle, we can assume that the *Period of the cycle* is 24.0 h (although you should always verify it first by running **Wintau.exe**). If the period is different from 24.0 h, there will be more (or less) than one full cycle in each 24 h, and the program must know about it in order to make the appropriate adjust-ments.

9. The *Number of cycles* panel allows you to specify how many cycles (which is the same as number of days if the period is 24.0 h) to use in the analysis. If you specify *One cycle*, only the first cycle will be loaded from the disk file. If you specify *More than one cycle*, you will be asked how many cycles should be used and how many cycles should be skipped before loading the cycles to be used (just in case the cycles you want to use are not at the beginning of the file). The various cycles are averaged into a single cycle. Let us choose *One cycle* first. Click on it.

10. In the unlikely case that you want to change the program's choice of higher and lower values, you can do so at the *Confirm parameters* panel. I recommend that you always just click on *OK*. Do it now.

11. The *Results* panel shows you the acrophase (which in this case is 1.7 h, or 1:42 am) and an index and interpretation of how good the cosine fit is (in this case, 0.11, which is a good fit). Although I have not told you so yet, the file started at midnight and the lights were on from 5 am to 7 pm. Thus, the acrophase is 3.3 h before lights on for this nocturnal rodent. The graph on the right allows you to subjectively access the acrophase and the goodness of fit.

12. Since the file contains 7 days of data, let us repeat the procedure using 7 days rather than just one day. Click on *Start a whole new analysis*.

13. Click on *Load data from disk*, then on *OK*, then on *File contains data only*, then on *OK*, then on *More than one cycle*. Click on the *Number of cycles to be averaged* box, delete the number 1 and type the number 7. Then, click on *OK*. At the *Confirm parameters* panel, click on *OK*.

14. Now the computed acrophase is 1.2 h (rather than 1.7 h), and the goodness-of-fit index is 0.04 (excellent rather than just good). This is natural, as a more accurate calculation of acrophase is possible when more data are available.

15. Click on *Same parameters, new file*. Select file **A17.dat**. This file is similar to the previous one except that the animal whose body temperature was recorded was a tree shrew (which is diurnal) rather than a fat-tailed gerbil.

16. Since we told the program to use the same parameters with a different file, we do not need to enter all the information again (the low and high values,which, of course, are not the same in the two files, are automatically replaced). Click on *OK* to see the results.

17. The computed acrophase is 11.8 h (11:48 am), which is 6.8 h after lights on. Not surprisingly, the body temperature rhythm of the diurnal tree

shrew peaks during the day, whereas the rhythm of the nocturnal gerbil peaked at night.

exercise 7: calculating circadian acrophase (continued)

Having practiced acrophase analysis with simple data files in the previous exercise, we will now deal with more complex data sets.

1. Start the program **Acro.exe.**
2. Choose *Load data from disk.*
3. Select the sub-folder *Data* by double-clicking on it.
4. Select the data file **A18.dat** by clicking on it. This file contains the records of locomotor activity of a thirteen-lined ground squirrel measured by telemetry every six minutes for one day under a 24-h light-dark cycle (lights on from 3 am to 5 pm). Each line of the file contains a time stamp (the clock time in decimal format) and the activity count.
5. Click on *OK* (or double-click on the file name).
6. Click on *File contains times and data.*
7. Leave the *Period of cycle* as 24 and choose *One cycle.*
8. A window will appear to remind you that, because files containing time stamps may have data collected at irregular intervals, averaging of successive cycles is not possible. This is not a problem for us, as our file contains only one cycle. Click on *OK.*
9. As usual, just click on *OK* at the *Confirm parameters* panel.
10. The *Results* panel indicates that the acrophase is 10.3 h (or 10:18 am). This is 7.3 h after lights on, which is consistent with the diurnal habit of this squirrel species.
11. Particularly when human subjects are involved, it is not uncommon for researchers to have a limited data set at their disposal. Measurements may not be feasible more often than once an hour and, even then, some measurements may be missing. Acrophase analysis is still possible, however. The data file **A19.dat** contains measurements of my own oral temperature taken at hourly intervals for 24 hours. I missed one measurement at 3 am and one at 2 pm. You can look at the data by loading the file into your word processor after you complete the next step.
12. Click on *Same parameters, new file.* Select data file **A19.dat** and click on *OK.* As for most people who wake up each day at approximately 7 am and go to sleep at 11 pm, the acrophase of my body temperature is 5 pm.
13. Minimize or close **Acro.exe** and load **A19.dat** into your word processor. As you can see, the structure of the file is rather simple. Each time point (in a 24-h clock) is paired with a measurement of body temperature. You can modify the file, save it under a different name (in text

format), and analyze it with **Acro.exe** later. For now, let us go back
to the program and practice manual data entry.

14. Restore or open **Acro.exe**. Click on *Enter data manually.*
15. Click on the top box under *Time units* and enter the following numbers
    (pressing the *ENTER* key or *Down Arrow* key after each one): 5, 10,
    16, 23, 0, 3. Then click on the top box under *Data values* and enter
    these numbers: 9, 6, 0, 3, 3, 5. When done, click on *OK*, and then *OK*
    again at *Confirm parameters* (which should be 0 and 9).
16. If you entered the data correctly, the *Results* panel should indicate an
    acrophase of 8 h with an excellent cosine fit.

simulating the circadian clock

exercise 8: building a phase-response curve

The program **Winmodel.exe** is a functional model of the mammalian circadian pacemaker with parameters set to simulate the circadian system of the
golden hamster. Many circadian processes, including free-run in constant
darkness or constant light, photic entrainment, and relative coordination,
can be simulated. In this exercise, we will use multiple light-induced phase-
shifts to build a phase-response curve.

1. Start the program **Winmodel.exe**.

2. At the top window, change *Period* to 23.5 h.
3. Set the number of *Blocks* to 2. Leave the *Natural* box unchecked.
4. At the Block 1 window, leave all values at their defaults (*Days* at 14,
   Mode at LD, *Lights on* at 0, and *Lights off* at 12, which means that the
   "animal" will be under a 24-h light-dark cycle, with lights on at
   midnight and off at noon, for 14 days).
5. At the Block 2 window, leave *Days* at 14, select the pulse *Mode*, and
   leave the *Days in DD* at 7 and *CT time* at 20. This means that the
   animal will be in constant darkness for 14 days, except for a light
   pulse at circadian time 20 on the eighth day. Circadian time 20 is 8
   circadian hours after the onset of activity (circadian time 12). Because
   the free-running period is different from 24.0 h, one circadian hour is
   $\tau/24$ of a real hour. The computer will make the necessary adjustment
   (for $\tau = 23.5$, one circadian hour equals 0.98 h).
6. At the top window, click on *RUN*.
7. You will observe a phase advance of 1.75 h after the light pulse.
   Since each physical hour is $24/\tau$ of a circadian hour, the phase shift
   is +1.79 h.
8. Since the conversion between physical hour and circadian hour can
   be confusing, let us construct the phase-response curve using an animal with a period of 24.0 h, so that one circadian hour is the same
   as one physical hour. At the top window, change the *Period* back to 24.

9. In the Block 2 window, change the *CT time* to 0. At the top window, click on *RUN*.
10. You should see no phase shift (a phase shift of 0.0 circadian hours). Write down the circadian time of the pulse and the resulting phase shift. Then, repeat step 9 eleven times, replacing only the *CT time* (2, 4, 6, ..., 22). You will observe phase delays at CT 12 and CT 14 and phase advances from CT 16 to CT 22.
11. When you have collected all twelve data points, plot them on a graph with circadian time on the abscissa and phase shifts on the ordinate. The graph should look similar to the hamster phase response curve shown in Figure 5 (Chapter 5).
12. If you are interested in seeing how the length of exposure to constant darkness affects the photic sensitivity of the circadian system, repeat the whole procedure with a few alterations in the Block 2 window: set *Days* to 21 and *Days in DD* to 14. The amplitude of the resulting phase-response curve should be much greater than the one obtained after only 7 days in DD.
13. Afterwards, you can repeat the procedure again with a longer duration of constant darkness, but the amplitude of the phase-response curve will not increase any further (saturation is achieved at 14 days). On the other hand, durations shorter than 7 days should reduce the amplitude of the curve.
14. At your discretion, you may wish to vary many of the parameters and compare free-runs in DD with free-runs in LL, phase angles of entrainment for different circadian periods, and so on. To make the plots look more "natural" (i.e., exhibiting random daily variations in onsets and offsets, brief interruptions in running-wheel activity, and transients after phase shifts), check the *Natural* box at the top window.

exercise 9: building a circadian clock

The program **Free-run.exe** allows you to simulate two particular phenomena: the pooling of SCN cells with varying circadian periods to produce a single "organismal" period, and the production of "temporal chimeras" by placing two distinct circadian pacemakers in the same animal (as in hamsters with partial SCN lesions receiving a hypothalamic graft from a genotypically distinct hamster). These two phenomena are described in Chapter 9.

1. Start the program **Free-run.exe**.
2. At the top panel, select *Two pacemakers*.
3. Set the *Period* and *Strength* of the two pacemakers at your discretion and click on *Plot*. For instance, if one pacemaker has a period of 24 h and a strength of 60%, while the other has a period of 20 h and a strength of 40%, the actogram of organismal activity will look like a series of horizontal lines dense at the center. If you look closely, you will be able to identify both the 24-h and the 20-h components.

4. Repeat the procedure with other combinations of period and strength. You may have a hard time identifying the 24-h component in the plot generated by a pacemaker with 24-h period and 40% strength combined with a pacemaker with 22-h period and 60% strength, for example.

5. After you have exhausted your curiosity, select *One pacemaker* at the top panel.

6. Let us start with all the default values, which are 1 for the *No. of cells*, 22 h for the *Lowest period*, and 26 h for the *Highest period*. Click on *Plot*.

7. The actogram shown will have a circadian period anywhere in the 22- to 26-h range. Click on *Plot* many times and observe the various periods.

8. Change the *No. of cells* to 100 and run several simulations again. Notice how the range of periods is greatly reduced. Since the period of the individual cells ranges from 22 to 26 h, the mean period of all 100 cells tends to 24.0 h.

9. Change the *No. of cells* to 1000 and run a few more simulations. Notice how the range of periods is reduced even further.

10. Because the *Integration site* was left at the default value (*Oscillator*), the reduction in the range of expressed periods was a natural consequence of the averaging process. That is, the period of the expressed rhythm was the mean of the periods of the various cells. Statisticians have long known that the standard deviation of a distribution of means is much smaller than the standard deviation of the distribution of the elements. However, it is theoretically possible that the integration of the period of individual SCN cells takes place not at the SNC itself (the "Oscillator") but at the organs that receive the output of SCN cells (the "Effectors"). What would happen if the effector organs responsible for locomotor activity received multiple inputs from SCN cells with different circadian periods?

11. Select *Effectors* as the *Integration site*. Unless you want to spend an hour waiting for the simulation to be completed, make sure to reduce the *No. of cells* to 2. Then click on *Plot*.

12. The outcome does not look very natural. Run the simulation a few more times. Notice that none of them looks like the record of locomotor activity of a normal animal.

13. Change the *No. of cells* to 10 and run several more simulations. The output looks even more unusual. It would seem that integration at the effector organs is *not* what happens in real animals. Of course, you have not tried a larger number of cells yet, but I can tell you that the output is still abnormal for 1000 cells. You can try it for yourself. If your computer is slower than 100 MHz, the simulation may take a quarter of an hour; at 400 MHz, it should take a couple of minutes. Change the *No. of cells* to 1000 and click on *Plot*.

minimizing jet lag

exercise 10: picking the best time to travel

This last exercise is meant for entertainment just as much as for education. You may want to share it with family members and friends.

1. Start the program **Jet-lag.exe**.
2. Click at the button under *Origin*, then click on Florida in the map (which should be bright blue in most monitors).
3. Click at the button under *Destination*, then click on Western Europe in the map (dark blue).
4. Because this trip takes you out of North America, a message is displayed to remind you that the beginning and end of Daylight Savings are not identical in other parts of the world. Thus, if you are traveling around Easter or Halloween, the time-zone differential may be an hour off. Click on the button under *Travel*.
5. A message is displayed telling you that you will cross 6 time zones eastward. The best way to avoid jet lag under this condition is to arrive late at night, go to bed late, and wake up late the next day.
6. Now, let us change the origin of the trip. Click at the button under *Origin*, then click on California (purple), then click on the button under *Travel*.
7. The program will tell you that you will cross 9 time zones to go from California to Western Europe. This will require a large shift of your internal clock. Although you could try the same approach as in the previous trip (item 5 above), most airlines give up on attempting to attain synchronization and simply try to optimize your ability to sleep on the first night at the destination. They have you travel overnight (and get very little sleep) and arrive early in the morning (local time). By the time the locals are going to bed, you have been awake for so long that you are likely to sleep reasonably well even though your internal clock says that it is not time to sleep.
8. Repeat the procedure with any origin and destination that you wish. If your travel requires you to cross the Equator, a message will be displayed to remind you that Daylight Savings, if applicable, are effective at opposite times of the year in the two hemispheres. Thus, the time-zone differential may be off by up to two hours. You should consult the consulate of the particular country to determine if and when Daylight Savings begin and end. All calculations provided by this program are based on standard time at both origin and destination.

bibliographic notes

1. Refinetti, R. (1996). Comparison of the body temperature rhythms of diurnal and nocturnal rodents. *Journal of Experimental Zoology* 275: 67-70.
2. Refinetti, R. (1992). Analysis of the circadian rhythm of body temperature. *Behavior Research Methods Instruments and Computers* 24: 28-36.
3. Nelson, W., Tong, Y. L., Lee, J. K., and Halberg, F. (1979). Methods for cosinor rhythmometry. *Chronobiologia* 6: 305-323.
4. Enright, J. T. (1965). The search for rhythmicity in biological time-series. *Journal of Theoretical Biology* 8: 426-468.
5. Dörrscheidt, G. J. and Beck, L. (1975). Advanced methods for evaluating characteristic parameters of circadian rhythms. *Journal of Mathematical Biology* 2: 107-121.
6. Sokolove, P. G. and Bushnell, W. N. (1978). The chi square periodogram: Its utility for analysis of circadian rhythms. *Journal of Theoretical Biology* 72: 131-160.
7. Bloomfield, P. (1976). *Fourier Analysis of Time Series: An Introduction.* New York: Wiley.
8. Gottman, J. M. (1981). *Time-Series Analysis: A Comprehensive Introduction for Social Scientists.* New York: Cambridge University Press.
9. Pittendrigh, C. S. and Daan, S. A. (1976). A functional analysis of circadian pacemakers in nocturnal rodents. I. Stability and lability of spontaneous frequency. *Journal of Comparative Physiology* 106: 233-252.
10. Albers, H. E., Gerall, A. A., and Axelson, J. F. (1981). Effect of reproductive state on circadian periodicity in the rat. *Physiology and Behavior* 26: 21-25.
11. Refinetti, R. (1991). An extremely simple procedure for the analysis of circadian and estrous periodicity. *Physiology and Behavior* 50: 655-659.
12. Refinetti, R. (1993). Comparison of six methods for the determination of the period of circadian rhythms. *Physiology and Behavior* 54: 869-875.
13. Refinetti, R. (1991). Use of chi square periodogram in the analysis of estrous rhythmicity. *International Journal of Biomedical Computing* 27: 125-132.
14. Ruf, T. (1999). The Lomb–Scargle periodogram in biological rhythm research: Analysis of incomplete and unequally spaced time-series. *Biological Rhythm Research* 30: 178-201.
15. Refinetti, R. (1992). Non-parametric procedures for the determination of phase markers of circadian rhythms. *International Journal of Biomedical Computing* 30: 49-56.
16. Ruf, T. (1996). The baseline cosinus function: A periodic regression model for biological rhythms. *Biological Rhythm Research* 27: 153-165.
17. Wang, Y. and Brown, M. B. (1996). A flexible model for human circadian rhythms. *Biometrics* 52: 588-596.

Index